Cognitive Radio-based Internet of Vehicles

The incorporation of Cognitive Radio (CR) into the Internet of Vehicles (IoV) has emerged as the Intelligent Transportation System (ITS). It covers the aspects of cognitive radio when it provides support to IoV and the challenges that limit the performance of ITS. These issues include unreliable delivery, the dynamic topology of IoV, routing overhead, scalability, and energy, to name a few. These challenges can be considered as future research directions for a promising intelligent transportation system. Machine learning (ML) is a promising discipline of Artificial Intelligence (AI) to train the CR-based IoV system. It also covers the applications of ML techniques to the CR-IoV systems and highlights their issues and challenges. Finally, it includes an examination of ML in conjunction with Data Science applications. In CR-IoV, ML and Data Science can be collaboratively used to further enhance road safety through inter-vehicle, intra-vehicle, and beyond-vehicle networks.

The channel switching and routing overhead is an important issue in CR-based IoVs. To minimize the channel switching and routing overheads, an effective scheme has been presented in Section 4 to discuss the promising solutions and performance analysis. Meanwhile, IoV communication is a time-sensitive application that requires vehicles to be synchronized. The time synchronization in IoVs has been highlighted in Section 5 to elaborate further on the critical metrics, challenges, and advancements in synchronization of IoVs. As the vehicles exchange data using wireless channels, they are at risk of being exposed to various security threats. The eavesdropping, identity exposure, message tampering, or sinkhole attack to name a few. It needs time to discuss the security issues and their countermeasures to make the CR-IoV attack resilient. The book concludes by addressing security issues and maintaining the quality of service (QoS) of the CR-based IoVs.

Key features:

- The architecture and applications of Intelligent Transportation System (ITS) in CR-IoVs.
- The overview of ML techniques and their applications in CR-IoVs.
- The ML applications in conjunction with Data Science in CR-IoVs.
- A minimized channel switching and routing (MCSR) technique to improve the performance of CR-IoVs.
- Data Science applications and approaches to improve the inter and intra-vehicle communications in CR-IoVs.
- The classification of security threats and their countermeasures in CR-IoVs.
- The QoS parameters and their impact on the performance of the CR-IoV ecosystem.

The target audience of this book is undergraduate and graduate-level students, researchers, scientists, academicians, and professionals in the industry. This book highlights future research directions that can be taken as research topics for future research.

Cognitive Radio-based Internet of Vehicles

Architectures, Applications and Open Issues

Edited by
Syed Hashim Raza Bukhari,
Muhammad Maaz Rehan, and
Mubashir Husain Rehmani

CRC Press is an imprint of the
Taylor & Francis Group, an informa business

Designed cover image: © Shutterstock

First edition published 2025
by CRC Press
2385 NW Executive Center Drive, Suite 320, Boca Raton FL 33431

and by CRC Press
4 Park Square, Milton Park, Abingdon, Oxon, OX14 4RN

CRC Press is an imprint of Taylor & Francis Group, LLC

ISBN: 978-1-032-25755-6 (hbk)
ISBN: 978-1-032-25759-4 (pbk)
ISBN: 978-1-003-28487-1 (ebk)

DOI: 10.1201/ 9781003284871

Typeset in Sabon
by Apex CoVantage, LLC

Contents

About the Editors

Syed Hashim Raza Bukhari (M'21, SM'21) is an associate professor in the Department of Electrical & Computer Engineering, Air University, Islamabad, Pakistan and College of Computer Science and Information Technology, King Faisal University, Saudi Arabia. He is the director of the Next Generation Intelligent Networks (NGIN) research group. He also served at COMSATS University Islamabad for nine years. Hashim obtained his Ph.D. in electrical engineering in 2017 from COMSATS University Islamabad (CUI), Wah Campus, Pakistan. Earlier, he received his M.S. and B.Eng. in computer engineering in 2011 and 2007, respectively. He has more than 15 years of experience in academics and has received numerous appreciations for his contributions to the improvement of teaching standards. He also received the research productivity award for his research contributions in 2017 from COMSATS. His research interests include issues in wireless sensor networks with dynamic spectrum access, cognitive radio networks, and ad hoc networks. Hashim is currently serving as Editor for *IEEE Future Direction Newsletter* and Guest Editor for *Journal of Networks and Systems Management* (Springer). He is also serving as a reviewer for *IEEE Communications Surveys and Tutorials, IEEE Communication Magazine, IEEE Transactions on Wireless Communications, IEEE Transactions on Vehicular Technology, IEEE Communication Letters, Pervasive and Mobile Computing* (Elsevier), *Computers and Electrical Engineering* (Elsevier), IEEE *Access, Wireless Networks* (Springer), *Future Generation Computer Systems* (Elsevier), and *Ad Hoc Sensor Wireless Networks.*

Personal Webpage: https://sites.google.com/view/hashimbukhari

Muhammad Maaz Rehan (M'16, SM'17) is a tenured associate professor in the Department of Computer Science, COMSATS University Islamabad, Wah Campus, Pakistan, and is leading the Intelligent SEcure Networks (IntelliSEN) Research Group. He obtained a Ph.D. from Universiti Teknologi PETRONAS (UTP), Malaysia, in January 2016 with two Bronze medals. He is twice an Internet Society (ISOC) fellow for the IETF 82nd and 87th, an ISOC fellow for Global INET 2012, an editor of IEEE Softwarization Newsletter, and an associate editor of *IEEE Access* journal. He is the senior lead in the Erasmus+ SAFE-RH Project, *Sensing, ArtiFicial intelligence, and Edge networking towards Rural Health monitoring* (project no. 619483-EPP-1–2020–1-UK-EPPKA2-CBHE-JP). Dr. Maaz is the lead editor of the book *Blockchain-Enabled Fog and Edge Computing: Concepts, Architectures and Applications*, published by Taylor & Francis Group, CRC Press, USA. The research areas of Dr. Maaz include blockchain and smart contracts, Internet of Everything (IoT/IoV), cybersecurity, and AI-based networks. Dr. Maaz has been involved in curriculum design and development activities and possesses a blend of extensive teaching experience at the national and international levels.

Personal Webpage: https://sites.google.com/view/maazrehan/

Mubashir Husain Rehmani (M'14-SM'15, SFHEA) received a B.Eng. degree in computer systems engineering from Mehran University of Engineering and Technology, Jamshoro, Pakistan, in 2004; an M.S. from the University of Paris XI, Paris, France, in 2008; and a Ph.D. from the University Pierre and Marie Curie, Paris, in 2011. He is currently working as Lecturer in the Department of Computer Science, Munster Technological University (MTU), Ireland. Prior to this, he worked as Post Doctoral Researcher at the Telecommunications Software and Systems Group (TSSG), Waterford Institute of Technology (WIT), Waterford, Ireland. He also served for five years as an assistant professor at COMSATS Institute of Information Technology, Wah Cantt., Pakistan. He is serving as an editorial board member of *NATURE Scientific Reports*. He is currently an area editor of *IEEE Communications Surveys and Tutorials*. He served for three years (from 2015 to 2017) as an associate editor of *IEEE Communications Surveys and Tutorials*. He served as Column Editor for Book Reviews in *IEEE Communications Magazine*. He is appointed as Associate Editor for *IEEE Transactions on Green Communication and Networking*. Currently, he serves as Associate Editor of *IEEE Communications Magazine*, Elsevier's *Journal of Network and Computer Applications (JNCA)*, and the *Journal of Communications and Networks (JCN)*. He is also serving as a guest editor of Elsevier's *Ad Hoc Networks*

journal, Elsevier *Future Generation Computer Systems* journal, the *IEEE Transactions on Industrial Informatics*, and Elsevier's *Pervasive and Mobile Computing* journal. He has authored/edited a total of eight books, two books with Springer; two books published by IGI Global, USA; three books published by CRC Press–Taylor and Francis Group, UK; and one book with Wiley, UK. He received the Best Researcher of the Year 2015 of COMSATS Wah award in 2015. He received the certificate of appreciation, Exemplary Editor of the IEEE Communications Surveys and Tutorials for the year 2015, from the IEEE Communications Society. He received the Best Paper Award from the IEEE ComSoc Technical Committee on Communications Systems Integration and Modeling (CSIM) in IEEE ICC 2017. He consecutively received a research productivity award in 2016–17 and also ranked #1 in all engineering disciplines from the Pakistan Council for Science and Technology (PCST), Government of Pakistan. He received the Best Paper Award in 2017 from the Higher Education Commission (HEC), Government of Pakistan. He was the recipient of the Best Paper Award in 2018 from the Elsevier *Journal of Network and Computer Applications*. He was the recipient of the Highly Cited Researcher award thrice, in 2020, 2021, and 2022, from Clarivate, USA. His performance in this context features in the top 1% of citations in the field of computer science and cross field in the Web of Science citation index. He is the only researcher from Ireland in the field of computer science who has received this prestigious international award. In October 2022, he received Science Foundation Ireland's CONNECT Centre's Education and Public Engagement (EPE) Award 2022 for his research outreach work and being a spokesperson for achieving a work-life balance for a career in research.

Personal Webpage: https://sites.google.com/site/mubrehmani/

Contributors

Farhan Aadil
COMSATS University Islamabad
Attock Campus
Pakistan

Amine Abouaomar
Department of Computer and Software Engineering
Polytechnique Montreal
Montreal (QC) Canada

Syed Hashim Raza Bukhari
Department of Electrical and Computer Engineering
Air University
Islamabad, Pakistan
College of Computer Science and Information Technology
King Faisal University
Saudi Arabia

Soumaya Cherkaoui
Department of Computer and Software Engineering
Polytechnique Montreal
Montreal (QC) Canada

Abderrahime Filali
Department of Computer and Software Engineering
Polytechnique Montreal
Montreal (QC) Canada

S. M. Usman Hashmi
Department of Computer Science
Bahria University
Islamabad Campus, Pakistan

Muntazir Hussain
Department of Electrical and Computer Engineering
Air University
Islamabad, Pakistan

Zeeshan Iqbal
University of Engineering and Technology Taxila
Pakistan

Ehsan Ullah Munir
Department of Computer Science
COMSATS University Islamabad
Wah Campus, Pakistan

Muhammad Nadeem
Department of Computer Science
COMSATS University Islamabad
Wah Campus, Pakistan

Mudassar Naseer
Department of Computer Science
The University of Lahore
Lahore, Pakistan

Muhammad Nouman Noor
Department of Computer Science
Hitec University Taxila
Pakistan

Ali Raza
University of Engineering and Technology Taxila
Pakistan

Muhammad Maaz Rehan
Department of Computer Science
COMSATS University Islamabad
Wah Campus, Pakistan

Saad Rehman
International Institute of Science, Art and Technology (IISAT)
Gujranwala, Pakistan

Saddaf Rubab
Department of Computer Engineering, College of Computing and Informatics
University of Sharjah

Afaf Taik
Universit´e de Sherbrooke
Shebrooke,(QC) Canada

Mirza Anas Wahid
Department of Software and IT Engineering
´Ecole de technologie sup´erieure
Montreal, Canada

Muhammad Naveed Younis
Department of Computer Science
The University of Lahore
Lahore, Pakistan

Muhammad Zeeshan
Walton Institute for Information and Communication Systems Science
Waterford, Ireland

Section 1

Introduction to CR-IoV

Chapter 1

Intelligent Transportation System in Cognitive Radio Internet of Vehicles

Ali Raza, Zeeshan Iqbal, Syed Hashim Raza Bukhari, and Farhan Aadil

1.1 INTRODUCTION

"Intelligent transportation system" (ITS) refers to the use of data collection, processing, and communication technologies in a transportation system [1]. It is considered one of the major building blocks of any smart city. There could be no smart cities without efficient and reliable ITSs. The core aim of ITSs is to increase the safety of people and vehicles, reducing traffic congestion, and managing accidents effectively while providing an entertaining driving and traveling experience. It is stated that around 90% of road accidents are caused by human errors and misjudgments [2]. This necessity of providing the safe and entertaining driving makes the ITS a key component of the smart city concept. In order to reduce the number accidents by effectively reducing the impact of human error, the International Telecommunication Union-Telecommunication (ITU-T) standardization sector proposed the concept of the vehicular ad-hoc network (VANET) in 2003. A VANET is the key enabling technology for successful implementation of ITSs. It provides a communication platform for vehicles and enables them to send/receive early warning messages. The warning message either warns the driver before they make a mistake or provides information about accidents or unusual events.

Despite having great potential, VANETs have long encountered multiple barriers in wide applications and have not brought great commercial value [3, 4]. These barriers include unstable network service quality, temporary networks with limited coverage, incompatibility with personal communication devices, and lack of ability to process big data.

In the meantime, the concept of the Internet of Things (IoT) has revolutionized the world. It brings intelligence and automation in the field of smart homes, smart cities, smart industries, healthcare, and agriculture, to name a few. In order to use the prominent features of the IoT in ITSs, including the IoT in VANET brings about the era of the Internet of Vehicles (IoV) [5, 6]. The IoV forms a heterogenous network and supports Vehicle to Everything (V2X) communication facilities. In V2X, vehicles share information and communicate with all entities that may affect the vehicle either directly or

DOI: 10.1201/9781003284871-2

indirectly, with the aim of reducing accidents and providing information-oriented services. V2X communication encompasses vehicle to vehicle (V2V), vehicle-to-infrastructure (V2I), vehicle-to-personal devices (V2P), vehicle-to-sensors (V2S), and vehicle-to-roadside unit (V2R) communication [4].

1.2 COGNITIVE RADIO IN THE IoV

Integrating multiple heterogenous communication technologies in ITSs makes the IoV a complex network. It provides a single coherent platform and enables mutual cooperation of different communication models (V2V, V2I, V2R, V2P, V2S). Meanwhile, the adoption of IoT devices is peaking, and it is stated there will be 30.9 billion IoT devices connected to the internet by 2025 [7]. As a result of this emergence, the adoption of IoT devices in the IoV will also increase exponentially, making the IoV a huge network. In order to fully exploit the potential of the IoV, it must also support service-oriented applications in addition to safety applications. Service-oriented applications provide entertainment to passengers/drivers and enable them to connect to the internet and play online games or watch videos.

Apart from the other issues that the IoV network will face, spectrum scarcity is the prominent one [8]. A huge network that must also support service applications requires enough bandwidth for their seamless operation. Network devices (IoT, sensors, and vehicles) on the IoV are able to use the industrial, scientific and medical (ISM) and WAVE bands. These free bands are already overcrowded, and merging future networks into these will result in bandwidth contention issues.

In order to cope with wideband requirements, introducing cognitive radio (CR) in the IoV provides a promising solution [9]. CR scans the entire spectrum (licensed and free bands) and opportunistically uses unused spectrum holes. CR performs frequency hopping into free holes without interfering with the primary user.

1.2.1 Cognitive Radio

Cognitive radio (CR) is an intelligent radio communication system which opportunistically adapts to the surrounding environment and efficiently utilizes spectrum resources. The key concept behind CR is to learn the transmission parameters from the surrounding environment and use the underutilized portion of the spectrum. The cognitive or learning process starts from sensing the nearby environment in order to obtain spectral state information about communication parameters and detect any free holes in the spectrum. Utilization of the spectrum can be enhanced by dynamically adapting transmission parameters such as transmission frequency and power, modulation scheme, and signal-to-noise ratio (SNR).

As IoV networks generally operate in the industrial scientific medical (ISM) band, with the immense increase in wireless devices, these wireless channels are getting congested and becoming unsuitable for time-critical and data-intensive applications. In contrast to these bands, licensed bands are used for very short durations of time, and their utilization factor varies from 15% to 85% [10]. Keeping in view spectrum scarcity and the intensity of the imbalance between licensed and ISM bands, the Federal Communication Commission (FCC) has authorized unlicensed users (called secondary users; SUs) to use the licensed band, as long as it will not cause interference for licensed users (called primary users; PUs).

Most of the spectrum is licensed, and the spectrum hole is the band of frequencies that may not be utilized by PUs at certain times and locations, as shown in Figure 1.1. SUs exploit the spectrum holes and temporarily use them without interfering with the communication of PUs. When PUs reoccupy a channel, SUs must vacate it immediately.

1.2.2 Cognitive Cycle

The job of CR is to support SUs in sensing multiple channels, locating spectrum holes, choosing the most suitable hole, coordinating with neighbors, and adapting to the current environment. Once PUs reclaim an occupied channel, SUs must release it and search for another vacant channel. This process of sensing the channel, analyzing all available channels, selecting the best one, and switching over to another channel is termed the cognitive lifecycle, and it is depicted in Figure 1.2. The major stages of the CR lifecycle are described as follows.

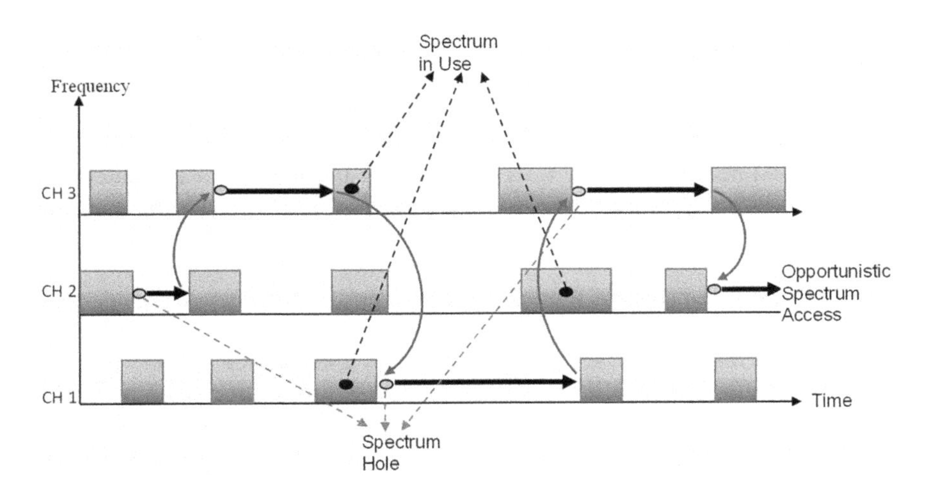

Figure 1.1 Dynamic spectrum access concept.

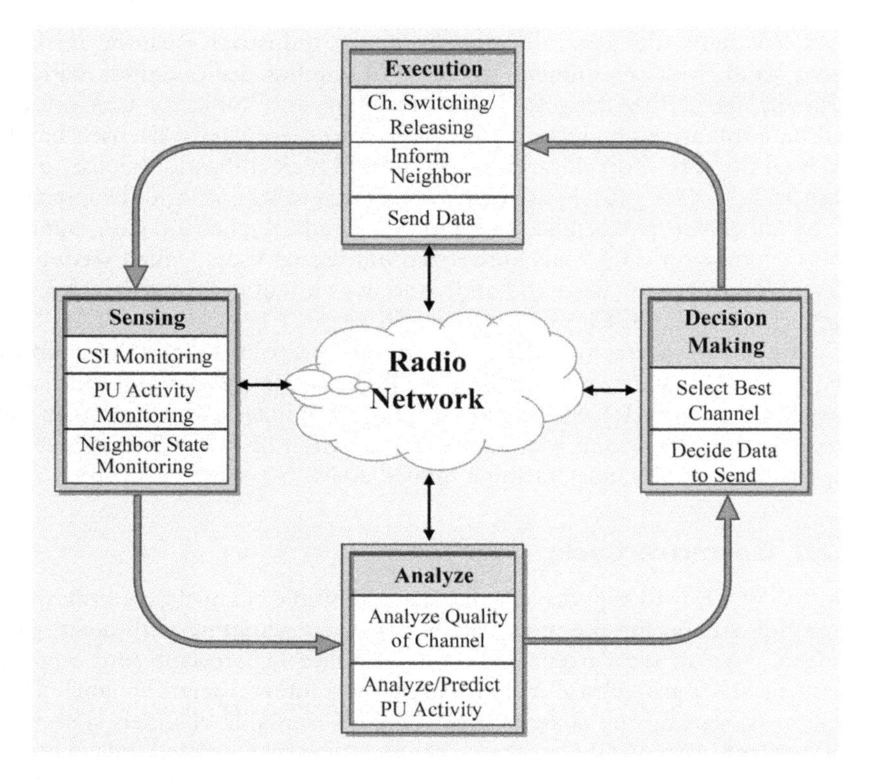

Figure 1.2 Cognitive radio lifecycle.

1.2.2.1 Spectrum Sensing

Spectrum sensing is the process of gathering channel state information (CSI), which includes channel characteristics, power, time and location of frequency in use, and local policies. Through spectrum sensing, a SU identifies the vacant frequencies or spectrum holes. Spectrum sensing also helps SUs obtain knowledge of the interference level and hence ensures that it will not harm PUs. There are three main approaches used to perform spectrum sensing: interference detection, cooperative sensing, and non-cooperative sensing.

The duration of sensing is considered an important parameter while sensing the spectrum. If the SU in CR senses for prolonged period of time, it will acquire more accurate information about a certain channel, and hence better utilization of the channel can be made without interfering with PUs. However, a long sensing duration wastes a significant amount of time and energy and is not suitable for battery-powered wireless CR nodes. Moreover, the long sensing duration puts an objection on its utilization for safety applications where delay is key concerns. On the other hand, sensing for a short duration of time saves time and energy but does not produce accurate information and results in frequent channel switching or interference with PUs. Therefore, it is required to select an optimal sensing duration that will give accurate CSI information and save time and energy as well.

1.2.2.2 Spectrum Analysis

Once spectrum sensing is performed, it gives CSI of available channels. Spectrum analysis includes rapid characterization of channels for the surrounding environment. PU activity on those channels is also analyzed, and a prediction is made about the next PU activity. Different machine learning algorithms can be used to characterize the channel and learn PU activity. This phase also includes negotiation with neighbor CR nodes to determine mutual cooperation and choose a suitable channel.

1.2.2.3 Spectrum Decision

Spectrum analysis helps to determine the best response strategy and choose the best channel for transmission out of available channels. The channel selection is based on PU activity, interference level, link error, path loss, and/or link delay.

1.2.2.4 Spectrum Mobility

This phase is also known as the execution phase. It involves switching over to the selected channel and starting transmission of data. It also involves continuing to observe the presence of PUs. As PUs appear on the selected channel, SUs must vacate it. In order to provide seamless communication, SUs must break the link and switch over to another vacant channel. Link breakage can also occur when CR nodes move out of each other's transmission range because of their mobility. The process of switching between different channels based on the network environment is called spectrum mobility. The spectrum mobility phase is followed by the sensing phase, and the process keeps on in the CR lifecycle.

1.3 APPLICATIONS OF CR-BASED IoV IN ITSs

The CR-assisted IoV provides a scalable solution for safety-related applications and wideband support for bandwidth-hungry service-oriented applications. The key purpose of safety applications of ITSs is to provide a safe journey to vehicles and passengers by delivering early warning messages to the driver about any special incidents. The warning messages help to extend the reaction time of driver so an accident can be avoided. Similarly, information from signs along the road also helps the driver foresee upcoming events.

The IoV in ITSs can also be used to control and manage traffic congestion on roads and provide navigational aid to the driver in order to complete the journey in a time and cost-effective manner. It can also be used to provide entertainment services to passengers, where they can play online videos/games while connecting with the onboard unit of vehicle, which will connect them to the internet. Some of the supporting applications of the CR-assisted IoV are shown in Figure 1.3. These applications are broadly classified into safety and service-oriented applications.

1.3.1 Safety

Nearly 1.25 million people die in road crashes each year, on average 3,287 deaths a day [11]. These numbers are even increasing day by day. The World Health Organization (WHO) is very concerned about such high death rates. With the use of ITSs (more specifically the IoV), these numbers can be reduced significantly. In the traditional collision avoidance system, the driver applies the brake or changes the road lane on visual observation of an obstacle or the brake lights of a front vehicle. The driver requires a reaction time of about 0.75 to 1.5 seconds [12] to perform safe actions. The prime reason to use the IoV in road safety applications is to provide the driver a longer reaction time on the occurrence of an event. If a driver is well aware of the surrounding vehicles, an early warning about the occurrence of an event may substantially reduce the number of accidents. On the occurrence

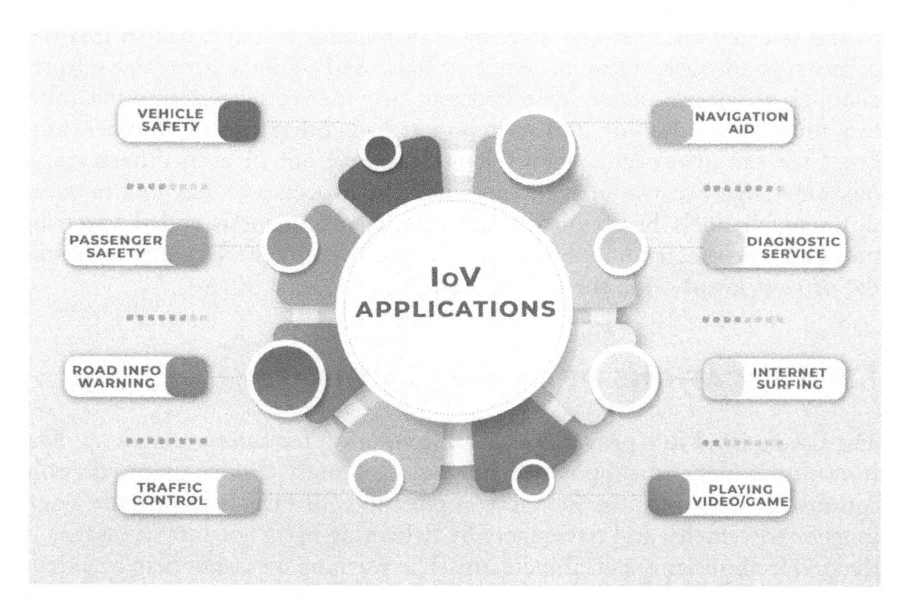

Figure 1.3 Safety and service-oriented applications of the CR-assisted IoV.

of an unusual event, the victim vehicle or the closet vehicle to the point of the incident informs other vehicles in close vicinity to the event. This provides enough time for other vehicles to react safely; hence, the accident can be avoided. The communication infrastructure of the IoV can also be used to assist drivers in lane changing and navigation; inform about speed limits, work zones, and post-crash notifications; and warn about road conditions and environmental hazards [13].

1.3.2 Service-Oriented Applications

With the successful implementation of ITSs, the IoV has also been adopted for traffic management, improving driving efficiency, and entertainment and user-oriented applications.

1.3.2.1 Traffic Management

Controlling and managing road traffic, especially in rush hours, helps to avoid traffic congestion/blocks and can significantly reduce traveling time and fuel consumption [14]. Vehicles periodically announce their positions, which is ultimately sent to a cloud server. The sever has a complete picture of the traffic of a town/city; it analyzes the traffic on each road and provides alternative paths to vehicles to reach their destination in a time- and cost-effective manner.

1.3.2.2 Navigational Aids

Apart from traffic management, drivers are assisted in order to facilitate and make their driving experience pleasant with help of 3D maps and augmented reality.

- 3D maps: Drivers may require a 3D map to visualize their surroundings. They might be searching for a restaurant, bank, or shopping mall. They may also visualize traffic on a particular road with the help of 3D maps.
- Augmented reality (AR): A considerable amount of research has been carried out on using AR technology in connected vehicles. With the support of AR, the driving and traveling experience can be made pleasant. AR also helps to avoid accidents by providing early warnings [15]. Three of the use cases of AR in connected vehicles are:

1. Traveling on the same road on a daily basis makes the outer view quite boring. AR aims to provide a visualization of a pleasant environment per the requirements of the user.
2. AR can help to provide visualization of blocked scenes, such as seeing beyond the vehicle in front (as shown in Figure 1.4).

Figure 1.4 Augmented reality visualization where it provides road-metery on head-up displays (bottom image) and view of a blocked scenes (top image).

3. AR-supported head-up displays provide an interactive view of outer environments, such as showing signboards; traffic signals; and the distance between front, side, and back vehicles.

1.3.2.3 Entertainment

People might be surfing the Internet, playing online 3D games, scrolling HD videos, or making a video call using their cellphones. These cellphones are connected with an in-vehicle mounted access point (AP), which is ultimately connected to the internet through a base station or roadside unit (RSU).

1.4 APPLICATION REQUIREMENTS

The long list of IoV applications has a diverse set of requirements. Some of them are delay sensitive, whereas others demand an ample amount of bandwidth to deliver large volumes of data. The focus of a few of them is providing guaranteed delivery in a highly dynamic environment, whereas others require quality of service (QoS) support with minimal overhead in a dense operating environment. Table 1.1 lists IoV applications and their requirements to satisfy user needs.

1.4.1 Delay

Timely delivery of safety/warning messages to the desired recipient(s) is the most important aspect of safety applications of the IoV [16]. These applications ask for delivery of information within the bounded delay so that the recipient(s) can take safe action to avoid an accident. If the message does not arrive in time, it may lead to severe consequences and can even cause fatal damages. A study in [15] attempted to increase driver reaction time by reducing the delay in message delivery, which results in reducing a significant number of vehicle collisions.

Passengers' infotainment and driver assistance with 3D maps and augmented reality services also aim for minimal delay for better user experience, but its impact is not critical, as in safety applications. These supporting applications can deliver user services in an effective manner with a moderate value of delay. In contrast to this, traffic management applications of the IoV are delay tolerant and have very little or no impact on user experience.

1.4.2 Guaranteed Delivery

Along with bounded delay, sure and guaranteed delivery of early warning messages is also a key pillar of safety applications on the IoV. Safety applications require a communication architecture where a message always arrives at the recipient, and there should be no packet collision along the path. Packet collisions trigger subsequent retransmission of packets that adds a significant amount of delay in delivery of messages and negates the purpose of the message. For broadcast messages, IoV applications also require the communication infrastructure to ensure guaranteed delivery to all recipients in the vicinity of event, and no vehicle should remain uninformed.

In contrast, non-safety applications have tolerance in the delivery of messages. These applications have elastic requirements in message delivery and can bear packet loss to some extent. They can deliver services with a bearable amount of loss without hampering user experience.

1.4.3 Bandwidth

In wireless communication, bandwidth is the width of the channel that will be occupied to transfer data over the medium. Shannon Hartley's theorem defines the relationship between bandwidth and amount of data transfer. The higher the bandwidth, the larger the amount of data that can be transferred with a given signal-to-noise ratio. It can be concluded that high bandwidth is required in applications where the user wishes to exchange a larger volume of data. In passenger infotainment applications, the user may be scrolling online videos, playing online games, or having video chats. All these services are data oriented and used to transfer huge amounts of data and hence require wide bandwidth for seamless services without interruption [17]. Bandwidth requirements become even more crucial for a scenario where bus passengers are surfing the internet via an access point situated in the bus. The access point must have a high-speed data connection with roadside units that are ultimately connected to a backbone network.

Augmented reality in vehicles is also used to transfer graphical and video content and therefore requires high bandwidth. In contrast to infotainment and augmented reality services, traffic management and safety applications do not put stringent requirements on bandwidth. Traffic management services can operate with an even narrower bandwidth, whereas safety applications may require moderate width to serve a congested vehicle environment. Since message transmission time is inversely proportional to the amount of bandwidth, safety applications can also use the benefits of a wide band in delivering warning information with reduced time delay.

1.5 ISSUES AND CHALLENGES

It is imperative to point out that in designing the CR-assisted Internet of Vehicles, attention must be paid to the particular characteristics of vehicular communication networks, such as the high speed of vehicles, high density in urban areas, and dynamic topology, which may lead to challenges in choosing a suitable network, communication channel, or data rate [18, 19]. In this section, various issues and challenges regarding implementation of the cognitive radio–assisted IoV in ITSs are discussed. These issues will be helpful in opening future research directions in the field of CR-assisted IoV.

Table 1.1 IoV Applications and their requirements

Applications		*Requirements*		
		Delay Sensitive	*Guaranteed Delivery*	*Bandwidth*
Safety	Vehicle/Passenger Safety	High	High	Medium
Service Oriented	Traffic Management	Low	Low	Low
	3D Maps and Augmented Reality	Medium	Medium	High
	Passenger Infotainment	Medium	Medium	High

1.5.1 Unreliable Delivery

The most important issue of the CR-assisted IoV is its inefficacy in providing guaranteed delivery [20]. The mobile nature of vehicles leads to unreliable V2V or V2X connections [21]. As vehicles move, the relative distance between nodes varies, which varies the propagation loss and hence causes the receiver-side signal indicator (RSSI) values at the receiving node to fluctuate. Moreover, buildings, trees, poles, and other vehicles also interrupt the radio waves and further worsen the effects of path loss. If the nodes are moving in opposite directions, RSSI variation occurs even more frequently, which results in an unreliable connection. Safety applications in ITSs require a highly reliable connection over which emergency messages can be delivered. Failure to receive emergency messages may lead to fatal accidents [22]. In addition, data-oriented services also require a reliable connection in order to achieve better throughput.

1.5.2 Dynamic Topology

The IoV includes RSUs, moving vehicles, and IoT devices, which act as mobile nodes. These mobile nodes may operate at extreme ranges of velocity and node density [23]. They may be stationary, positioned in parking, or struck in a traffic jam. They may be moving with a velocity of 150 km/h in a highway scenario. In the same way, vehicle density approaches the maximum at road cross-sections or traffic jams, especially in rush hours, whereas vehicles are sparsely located in highway scenarios. This heterogeneity of the IoV creates several issues. The high velocity of the nodes changes the network topology dynamically. Nodes join and leave the network frequently and cause increased communication overhead. Moreover, for high relative velocity, the IoV has to cope with the doppler effect, wastage of network bandwidth, frequent link failures, and high end-to-end delay.

1.5.3 Routing Overhead

The mobility of vehicles raises two critical issues as well: inadequate routing and frequent handover [24, 25]. In V2V architecture, vehicular nodes communicate in a multi-hop manner. For the sake of sending data packets in multiple hops, nodes maintain routes either proactively or reactively. The high mobility of nodes changes the position of the source, destination, and/or intermediate nodes and in turn interrupts the data path. This interruption initiates the route maintenance process, which is purely routing overhead. Increasing node mobility ultimately increases the routing overhead and hence lowers network performance [26]. The V2I communication mechanism also suffers from node mobility. It causes frequent handover among roadside units or base stations (BSs), leads to service interruption, and introduces additional delays.

1.5.4 Scalability

Scalability is also a major issue on the Internet of Vehicles [27]. If not handled effectively, increasing the number of nodes further downgrades network performance. The network becomes congested and adds a noteworthy amount of delay, causes excessive packet collisions, and enlarges the routing table.

1.5.5 Energy

Energy consumption during wireless communication is also a key component to evaluate the performance of the IoV. Although it is not given much importance in the literature for VANET communication, where energy comes from vehicle power, it can't be neglected for the IoV, which includes IoT devices with very limited battery energy [28, 29]. It can also be considered an essential parameter for battery-powered electrical vehicles [30, 31]. It also plays a role from the perspective of green computing [32].

1.5.6 Spectral Efficiency

The spectrum is a key scarce resource of wireless communication in general, particularly in the IoV. The width of the available spectrum determines the

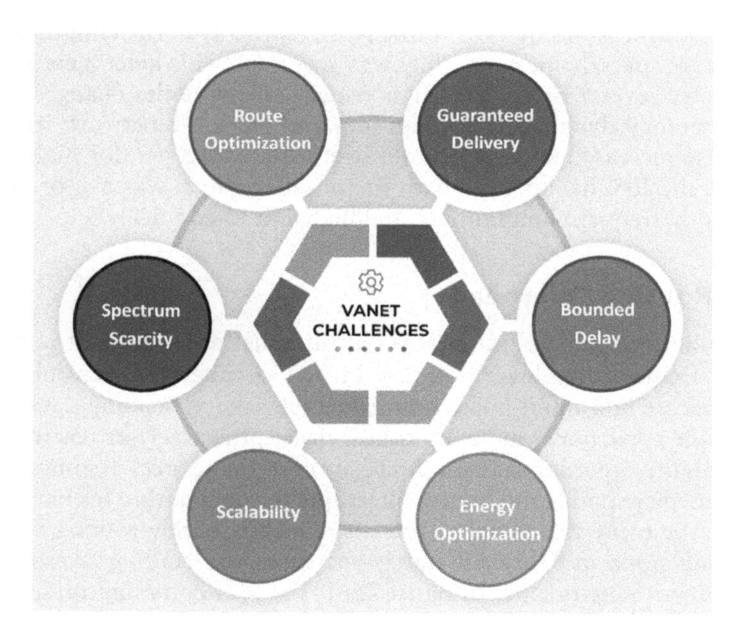

Figure 1.5 Issues and challenges encountered by VANET to fulfill application requirements.

amount of data that will be transferred. The wider the band of the available spectrum, the larger the amount of data that can be transferred in a given time. Although the integration of CR with the IoV aims to resolve the spectrum scarcity issue, significant efforts are required for its successful implementation [33, 34]. For instance, research on opportunistic utilization of licensee bands must make sure that there is no interference with PUs. Researchers should also pay attention to the spectrum decision and mobility phase in order to minimize the overhead while switching from one channel to another.

1.6 CONCLUSION

In this chapter, we have provided the motivation to integrate cognitive radio in the IoV. We highlighted the potential of cognitive radio and discussed the stages of the cognitive cycle. We have discussed that the CR-assisted IoV can be a good approach to meet the requirements of safety as well as service-oriented applications in smart cities. We concluded the discussion with possible research directions by presenting open issues which need to be addressed for the cognitive radio–assisted IoV in ITSs.

REFERENCES

1. G. Dimitrakopoulos and P. Demestichas, "Intelligent transportation systems," *IEEE Vehicular Technology Magazine*, vol. 5, no. 1, pp. 77–84, 2010.
2. N. A. Stanton and P. M. Salmon, "Human error taxonomies applied to driving: A generic driver error taxonomy and its implications for intelligent transport systems," *Safety Science*, vol. 47, no. 2, pp. 227–237, 2009.
3. C. Bhanu, "Challenges, benefits and issues: Future emerging VANETs and cloud approaches," *Cloud and IoT-Based Vehicular Ad Hoc Networks*, pp. 233–267, 2021.
4. B. Ji *et al.*, "Survey on the Internet of Vehicles: Network architectures and applications," *IEEE Communications Standards Magazine*, vol. 4, no. 1, pp. 34–41, 2020.
5. S. K. Indu, "Internet of Vehicles (IoV): Evolution, architecture, security issues and trust aspects," *International Journal of Recent Technology and Engineering*, vol. 7, no. 6, p. 2019, 2019.
6. A. Hakimi, K. M. Yusof, M. A. Azizan, M. A. A. Azman, and S. M. Hussain, "A survey on Internet of Vehicle (IoV): Applications & comparison of VANETs, IoV and SDN-IoV," *ELEKTRIKA-Journal of Electrical Engineering*, vol. 20, no. 3, pp. 26–31, 2021.
7. Statista, "Internet of Things (IoT) and non-IoT active device connections worldwide from 2010 to 2025." www.statista.com/statistics/1101442/iot-number-of-connected-devices-worldwide/ (accessed 30 December 2021).

8. J. Eze, S. Zhang, E. Liu, E. E. Chinedum, and Q. Y. Hong, "Cognitive radio aided Internet of Vehicles (IoVs) for improved spectrum resource allocation," in *2015 IEEE International Conference on Computer and Information Technology; Ubiquitous Computing and Communications; Dependable, Autonomic and Secure Computing; Pervasive Intelligence and Computing*, 2015: IEEE, pp. 2346–2352.
9. K. F. Hasan, T. Kaur, M. M. Hasan, and Y. Feng, "Cognitive Internet of Vehicles: Motivation, layered architecture and security issues," in *2019 International Conference on Sustainable Technologies for Industry 4.0 (STI)*, 2019: IEEE, pp. 1–6.
10. A. Bradai, T. Ahmed, and A. Benslimane, "ViCoV: Efficient video streaming for cognitive radio VANET," *Vehicular Communications*, vol. 1, no. 3, pp. 105–122, 2014.
11. WHO, "Annual global road crash statistics, 2019." www.who.int/news-room/fact-sheets/detail/road-traffic-injuries (accessed June 2020).
12. K. A. Brookhuis, D. De Waard, and W. H. Janssen, "Behavioural impacts of advanced driver assistance systems – An overview," *European Journal of Transport and Infrastructure Research*, vol. 1, no. 3, 2019.
13. A. Rasheed, S. Gillani, S. Ajmal, and A. Qayyum, "Vehicular ad hoc network (VANET): A survey, challenges, and applications," in *Vehicular Ad-Hoc Networks for Smart Cities*: Springer, 2017, pp. 39–51.
14. A. Thomas, G. Singal, and R. Kushwaha, "Smart vehicles for traffic management and systems using cloud computing," in *Vehicular Cloud Computing for Traffic Management and Systems*: IGI Global, 2018, pp. 178–199.
15. T. Taleb, A. Benslimane, and K. B. Letaief, "Toward an effective risk-conscious and collaborative vehicular collision avoidance system," *IEEE Transactions on Vehicular Technology*, vol. 59, no. 3, pp. 1474–1486, 2010.
16. A. Bujari, M. Conti, C. De Francesco, and C. E. Palazzi, "Fast multi-hop broadcast of alert messages in VANETs: An analytical model," *Ad Hoc Networks*, vol. 82, pp. 126–133, 2019.
17. K. N. Qureshi, F. Bashir, and N. U. Islam, "Link aware high data transmission approach for Internet of Vehicles," in *2019 2nd International Conference on Computer Applications & Information Security (ICCAIS)*, 2019: IEEE, pp. 1–5.
18. W. Shi, H. Zhou, J. Li, W. Xu, N. Zhang, and X. Shen, "Drone assisted vehicular networks: Architecture, challenges and opportunities," *IEEE Network*, vol. 32, no. 3, pp. 130–137, 2018.
19. S. Jaap, M. Bechler, and L. Wolf, "Evaluation of routing protocols for vehicular ad hoc networks in typical road traffic scenarios," *Proceedings of the 11th EUNICE Open European Summer School on Networked Applications*, pp. 584–602, 2005.
20. E. C. Eze, S.-J. Zhang, E.-J. Liu, and J. C. Eze, "Advances in vehicular ad-hoc networks (VANETs): Challenges and road-map for future development," *International Journal of Automation and Computing*, vol. 13, no. 1, pp. 1–18, 2016.
21. X. Ma, J. Zhang, X. Yin, and K. S. Trivedi, "Design and analysis of a robust broadcast scheme for VANET safety-related services," *IEEE Transactions on Vehicular Technology*, vol. 61, no. 1, pp. 46–61, 2011.
22. W. Li, W. Song, Q. Lu, and C. Yue, "Reliable congestion control mechanism for safety applications in urban VANETs," *Ad Hoc Networks*, vol. 98, p. 102033, 2020.

23. O. Senouci, Z. Aliouat, and S. Harous, "MCA-V2I: A multi-hop clustering approach over vehicle-to-internet communication for improving VANETs performances," *Future Generation Computer Systems*, vol. 96, pp. 309–323, 2019.
24. Y. Tang, N. Cheng, W. Wu, M. Wang, Y. Dai, and X. Shen, "Delay-minimization routing for heterogeneous VANETs with machine learning based mobility prediction," *IEEE Transactions on Vehicular Technology*, vol. 68, no. 4, pp. 3967–3979, 2019.
25. J. Pereira, L. Ricardo, M. Luís, C. Senna, and S. Sargento, "Assessing the reliability of fog computing for smart mobility applications in VANETs," *Future Generation Computer Systems*, vol. 94, pp. 317–332, 2019.
26. R. A. Santos, A. Edwards, R. Edwards, and N. L. Seed, "Performance evaluation of routing protocols in vehicular ad-hoc networks," *International Journal of Ad Hoc and Ubiquitous Computing*, vol. 1, no. 1–2, pp. 80–91, 2005.
27. J. Liu, J. Wan, B. Zeng, Q. Wang, H. Song, and M. Qiu, "A scalable and quick-response software defined vehicular network assisted by mobile edge computing," *IEEE Communications Magazine*, vol. 55, no. 7, pp. 94–100, 2017.
28. M. Elhoseny and K. Shankar, "Energy efficient optimal routing for communication in VANETs via clustering model," in *Emerging Technologies for Connected Internet of Vehicles and Intelligent Transportation System Networks*: Springer, 2020, pp. 1–14.
29. M. Patra, R. Thakur, and C. S. R. Murthy, "Improving delay and energy efficiency of vehicular networks using mobile femto access points," *IEEE Transactions on Vehicular Technology*, vol. 66, no. 2, pp. 1496–1505, 2016.
30. B. Luin, S. Petelin, and F. Al-Mansour, "Microsimulation of electric vehicle energy consumption," *Energy*, vol. 174, pp. 24–32, 2019.
31. Y. Xie *et al.*, "Microsimulation of electric vehicle energy consumption and driving range," *Applied Energy*, vol. 267, p. 115081, 2020.
32. A. Siddiqa, F. F. Qureshi, M. A. Shah, R. Iqbal, A. Wahid, and V. Chang, "CCN: A novel energy efficient greedy routing protocol for green computing," *Concurrency and Computation: Practice and Experience*, vol. 31, no. 23, p. e4461, 2019.
33. F. Yang, J.-H. Han, X. Ding, Z. Wei, and X. Bi, "Spectral efficiency optimization and interference management for multi-hop D2D communications in VANETs," *IEEE Transactions on Vehicular Technology*, vol. 69, no. 6, pp. 6422–6436, 2020.
34. H. Kour, R. K. Jha, and S. Jain, "A comprehensive survey on spectrum sharing: Architecture, energy efficiency and security issues," *Journal of Network and Computer Applications*, vol. 103, pp. 29–57, 2018.

Section 2

Machine Learning in CR-IoV

Chapter 2

Machine Learning Applications in CR-IoV

*Amine Abouaomar, Abderrahime Filali, Afaf Taik, and Soumaya Cherkaoui**

2.1 INTRODUCTION

The Internet of Vehicles (IoV) is considered an evolution of the vehicular ad-hoc network (VANET), which connects vehicles to environmental entities, including pedestrian devices, intelligent road infrastructure, other vehicles, and sensors [1]. Therefore, the IoV is expected to provide a plethora of services, including advanced road safety, traffic efficiency, and infotainment services to vehicles. These services are characterized by diversified requirements in terms of latency, reliability, and data rate. Yet the high density of IoV devices in the network leads to the spectrum scarcity problem. Cognitive radio (CR) technology [2] is an IoV key enabler ensuring efficient utilization of spectrum resources and highly reliable communication between IoV devices [3]. Indeed, IoV devices assisted by cognitive capabilities can achieve seamless connectivity to the IoV network by adapting data transmission and reception to the radio environment. Although the CR-based IoV enables a certain degree of intelligence in spectrum sharing and the IoV environment characteristics, it requires additional mechanisms for more flexible decision-making. For instance, the dynamic topology of the IoV network and the high speed of vehicles, alongside the varying requirements of IoV services, necessitate a more dynamic exploitation of spectrum resources to meet the quality of service (QoS) of IoV applications [4].

Machine learning (ML) techniques have been proven efficient in several areas by providing systems the ability to learn from collected data or from their experiences to make relevant and timely decisions [5]. In particular, ML presents great potential in solving complex and dynamic environment problems such as the CR-IoV environment through a versatile set of tools and algorithms. For instance, CR-IoV can benefit from ML to enhance both quality of service and quality of experience (QoE). In such a field of applications, ML is expected to provide efficient schemes for the CR-IoV to reduce latency-related aspects during both the cognitive cycle (e.g., reducing sensing time) and operational latency (e.g., delivery service latency) [3]. As spectrum sensing is a crucial component in CR, ML enhances it with dynamic adaptation to the IoV network environment from different perspectives,

DOI: 10.1201/9781003284871-4

namely high to extreme mobility, spectrum security, and network overhead and latencies. Another application of ML in the CR-IoV is the routing and load balancing of vehicular network traffic. Due to the challenging aspect of high mobility, ML in this case offers enhanced decision making regarding the optimal path for packets, load balancing of the allocated spectrum, and migration of different services. The CR-IoV can also leverage federated learning (FL) as a privacy-preserving alternative to sensor data sharing, as it consists of collaboratively training ML models and only exchanging model parameters [6, 7]. From the resource provisioning point of view, ML will empower the energy and spectrum efficiency at different stages, including spectrum sensing, sharing, and management. Indeed, the gathered data on the status and utilization of the spectrum as well as the day-to-day road traffic data would be used by ML models to assist in decision making concerning traffic management. CR technology may be utilized to provide the necessary spectrum to exchange massive data between autonomous vehicles through the network infrastructure. Moreover, security of the CR-IoV is crucial due to the sensitivity of the use case that involves human safety. Therefore, it is important to prevent malicious data providers (spectrum sensors) from emitting fake/faulty data. ML can improve security in the CR-IoV through detecting and preventing jamming attacks, distributed denial-of-service (DDoS) attacks, and identity emulation (e.g., a secondary user pretending to be the primary user). Finally, ML algorithms can be leveraged to improve infotainment applications in the CR-IoV through intelligent channel selection for content transfer.

Despite existing efforts in applying ML to the CR-IoV, the path toward full ML integration into the CR-IoV still presents many challenges [8]. From the CR perspective, it is crucial to investigate advanced spectrum sensing and explore roads towards other radio access technologies (e.g., Wi-Fi, 5G, and beyond) [9]. From the ML perspective, reinforcement learning (RL)-based algorithms are particularly hard to adopt in practice due to the slow learning process and the difficulty of accurately capturing several scenarios in simulated environments. In fact, RL-based techniques require an agent to learn everything through the exploration of new states through repeated interactions with the environment, hence spending a considerable amount of time in the learning process. Additionally, the CR-IoV needs to adapt to changing topologies, which often requires retraining ML models or even applying core changes to them.

In general, the goal of this chapter is to provide an overview of recent efforts in using ML for the CR-IoV and its applications, as well as presenting open challenges and future research directions. The objectives of this chapter can be summarized as follows:

- Providing a comprehensive overview of using CR in the IoV.
- Overviewing various techniques and enablers from ML in CR and providing guidelines to leverage them in the IoV.

- Discussing applications and efforts of ML in CR-IoV, specifically in spectrum sensing, road safety, traffic congestion, resource provisioning, routing, and infotainment content delivery.
- Identifying issues that currently exist in ML applications for the CR-IoV and highlighting future challenges and important future directions for CR-IoV ML applications.

In this chapter, we present an overview of CR networks in the IoV and their key enablers. Then, we present the main benefits of ML for CR and the way to leverage them in the IoV environment. We will also discuss relevant literature efforts on ML applications for the CR-IoV. Finally, we will provide current challenges in CR-IoV–enabled ML, and we provide a set of future research directions.

2.2 OVERVIEW OF ML TECHNIQUES

Machine learning refers to a subset of artificial intelligence techniques allowing creation of systems able to learn useful information and necessary knowledge for decision making through learning patterns from collected data and experience. The rise of popularity of ML in recent years has mainly been in view of the high efficiency of deep learning (DL) models in a wide set of applications requiring analysis of high-dimensional data such as computer vision and natural language processing. Figure 2.1 summarizes different ML techniques applied to enhance CR-IoV performance.

2.2.1 Categorization of ML Techniques

The set of ML techniques is threefold: supervised, unsupervised, and reinforcement learning. In addition to this categorization, in the following, we also present artificial neural networks (ANNs), which are key ML models used in the different categories of ML techniques.

2.2.2 Supervised Learning

The premise of supervised learning is that data are composed of features and labels. A training sample X has a label Y, and the goal of the model is to learn the correct mapping between X and Y, $Y = f(X)$. Supervised learning tasks can be categorized into classification and regression.

Classification tasks include applications like handwriting recognition, object detection, and sentiment analysis. In such tasks, Y values are discrete, as they represent classes or categories. Several ML algorithms were developed for classification tasks, such as K-nearest neighbors (KNN) and support vector machine (SVM). Regression algorithms are used in continuous space, and the variable Y is continuous. Examples include weather

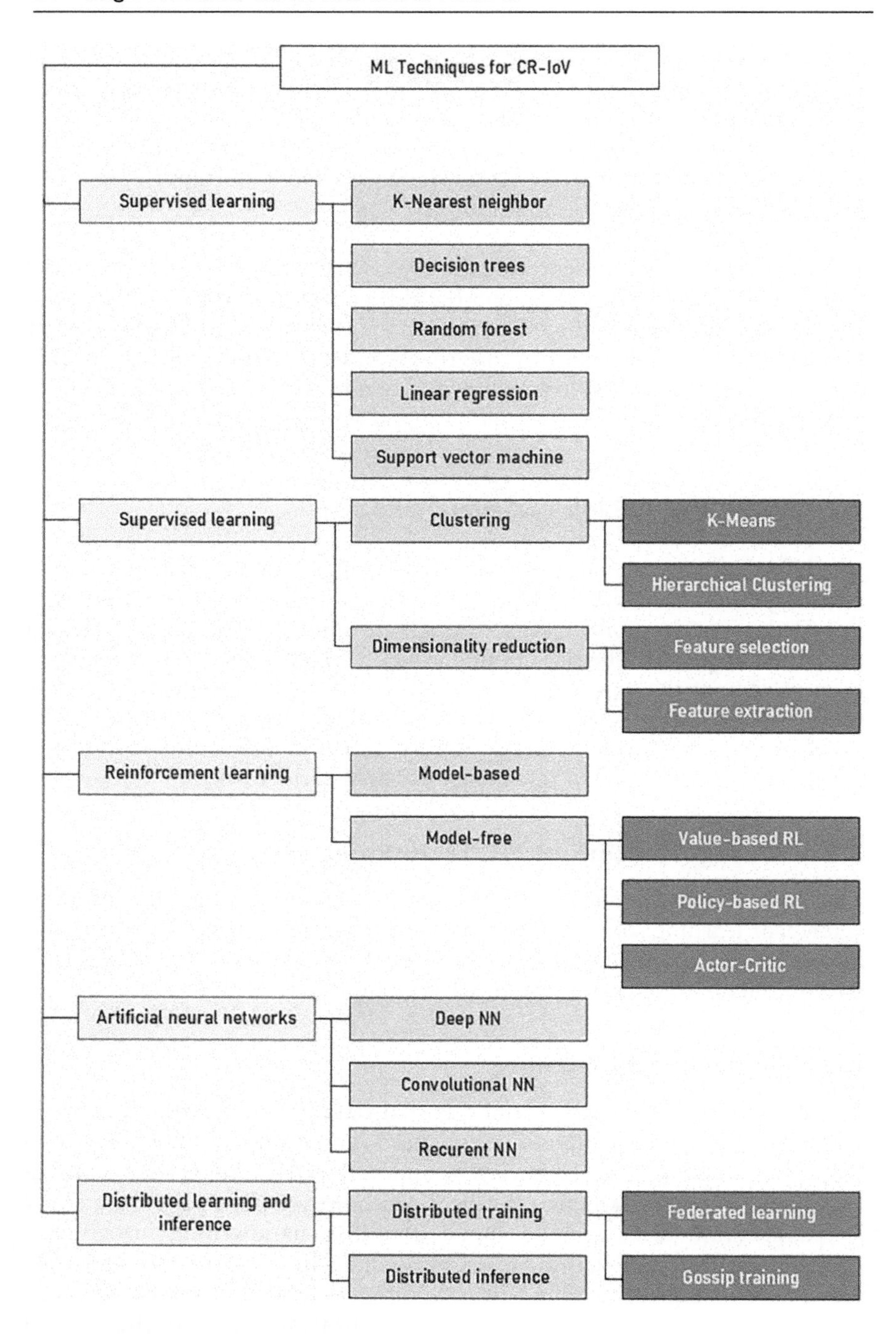

Figure 2.1 Machine learning approaches adopted in the CR-IoV.

forecasting, risk in finance, and drug response modeling. ML algorithms used for regression include linear regression and support vector regression (SVR). Other categories of algorithms can be used for both classification and regression, such as decision trees (DT), random forests, and artificial neural networks.

***K*-Nearest Neighbor:** KNN [**10**] is a simple classification algorithm where a data point's class is determined based on the distance of its feature values from those of other data points. The data point's class is then determined as the class of its *K* nearest neighbors. While the algorithm is easy to implement, robust against outliers, and suitable for multiclass classifications, it is very resource hungry when used with large datasets, as it requires calculating the distance from all the existing points.

Decision Trees: The goal of DTs is to learn simple decision rules inferred from the data features to predict a variable [**11**]. In a DT, each node represents a feature, each branch represents the conjunction of features that lead to classification, and each leaf node represents a specific class. The DT's training aims to maximize the information gain of each variable split, thus creating a hierarchy of the different features. DTs are simple to implement and result in explainable models. Nonetheless, they are prone to overfitting.

Random Forest [12], which usually consists of multiple DTs, randomly selects a subset of features to construct each DT. For classification tasks, the output of the random forest is the class selected by the most trees, while for regression, the predicted value is the mean or average prediction of the individual trees.

Linear Regression: The most simple form of regression is linear regression, where the goal is to fit a straight line to the dataset when the relationship between the variables of the dataset is linear. However, it oversimplifies real-world problems and is often unsuitable. In fact, linear regression models are classically obtained using the least squares approach, but to overcome its shortcomings, other variations were proposed, such as lasso [**13**] and ridge [**14**], which come with additional penalty parameters aiming to minimize the complexity and/or reduce the number of features used in the final model.

Support Vector Machine: The objective of SVM is to find a best-separating hyperplane in the feature space that maximizes the margin between two classes [**15**]. SVM is characterized by the use of different kernel functions such as linear, polynomial, and radial, where the choice of the function is crucial for classification accuracy. In general, SVMs have high stability and can achieve a low false alarm rate for binary classification tasks. On the downside, SVMs do not directly provide probability estimates, and they have high requirements in terms of memory and training time. In regression problems, a variant of SVM, support vector regression, aims to find a hyperplane that will fit the data points. SVR offers a flexible definition for the acceptable error threshold and requires little pre-processing of the data. Nonetheless, similarly to SVM, the choice of kernel plays a crucial role in SVR modeling and highly affects its performance.

2.2.2.1 *Unsupervised Learning*

Supervised learning algorithms are premised on the availability of large amounts of labeled data. However, labeling data is an expensive operation, as it requires a human or oracle's intervention, and is often deemed unfeasible. As a result, unsupervised learning algorithms can be used to remedy this challenge. In unsupervised learning algorithms, data points are uncategorized in nature, and the goal is to find patterns and similarities. Such algorithms can be categorized into clustering and dimensionality reduction. Clustering algorithms aim to segregate data samples into clusters by grouping the most similar data points. Meanwhile, dimensionality reduction aims to transform data from a high-dimensional space into a low-dimensional space so that the retained features have more representative properties from the original data, thus enhancing the performance of other learning algorithms in terms of memory usage and required computation.

Clustering: Several algorithms were developed for clustering, such as k-means, hierarchical clustering, spectrum clustering, and the Dirichlet process.

K-Means: The main idea behind k-means is to find k clusters in data by grouping the most similar data points with each other. The initial mean data points are chosen at random, and the new ones are calculated based on the feature similarity throughout iterations [**16**]. However, choosing the number of clusters to form (the value of k) is challenging, and the cluster formation quality is sensitive to the initial cluster centroids.

Hierarchical Clustering: Unlike k-means, which builds flat-clusters, hierarchical clustering builds a hierarchy of clusters, which are often represented as a dendrogram [**17**]. The clusters can be built in a bottom-up (i.e., agglomerative) or top-down (i.e., divisive) approach. The agglomerative approach starts with each data point as a standalone cluster and merges data points to build clusters, while the divisive approach begins with a single cluster and splits it using dissimilarity measures. While deciding the number of clusters to keep can be done through visualization, this choice might not be practical in systems requiring automation. Furthermore, hierarchical clustering is more computationally extensive compared to k-means.

Dimensionality Reduction: Dimension reduction is used in several applications, but it is especially useful for data summarization and compression. The algorithms that it utilizes can be categorized in different manners, including linear and non-linear. In this chapter, we consider two categories: feature selection [**18**] and feature extraction [**19**].

Feature Selection: Feature selection refers to a set of greedy algorithms that automatically select the subset of features most relevant to the learning problem. Such algorithms include sequential backward selection (SBS) [20] and the use of DT and its variants. SBS is a classical feature selection algorithm, which consists of iteratively eliminating features that have the minimum impact on a model. SBS can improve the quality of models in the case of overfitting.

Feature Extraction: In addition to a reduced number of features, feature extraction also applies transformations to the features. Notable algorithms used for feature extraction include principal component analysis (PCA), random projection, and independent component analysis (ICA). PCA detects the directions of maximum variance in high-dimensional data and projects the data points into a new subspace with the same or fewer dimensions. PCA consists of calculating the principal components, which are eigenvectors of the data's covariance matrix, and using them to perform a change of basis on the data. Nonetheless, PCA is not computationally heavy and may not be able to handle large datasets. Random projection [**21**] is a computationally efficient alternative to PCA, as it trades a controlled amount of error for faster processing and smaller models. Similarly to PCA and random projection, ICA [22] identifies the most important features of the original set. While PCA's main idea is to maximize variance, ICA's approach supposes that the features are mixtures of non-Gaussian independent sources, and its goal is to isolate these independent sources that are mixed in the dataset [**23**].

2.2.2.2 Reinforcement Learning

In reinforcement learning, agents learn to make decisions through interactions with their environment. The agent is in a game-like situation where, through trial and error, it learns to choose better actions to maximize the total reward. As RL is based on the feedback from the environment, it is highly suitable for real-time applications in dynamic and stochastic environments. As a result, RL was widely used for CR and IoV. In these applications, the agent can be a CR-based vehicle interacting with the radio environment. The agent might have enough knowledge about the environment to form a model about it in the case of model-based RL, as it might also not be able to form such a model and hence use model-free algorithms.

In the following, we consider an agent that interacts with an environment to learn an optimal policy π. In each time step t, the agent observes a state st S, where S is the set of possible states, and takes an action at A, with A the set of possible actions. As a result of the chosen action, the agent achieves a reward rt, and the environment transits to a new state st_{+1}.

Model-Based RL: Model-based RL consists of agents aiming to understand the environment and creating a model to represent it. In RL, the problem is often mathematically formulated as a Markov decision process (MDP). A MDP is a way of representing the dynamics of the environment, that is, the way the environment will react to the possible actions the agent might take at a given state.

In RL, the learning problem is often formulated as a Markov decision process [**24**] as a means to represent the way the environment will possibly react to the agent's possible actions. Such a reaction is modeled through a transition function, which takes the current state of the environment and the action as input and outputs a probability of moving to other states. The

MDP also includes a reward function, which represents the environment's response to a possible action. The reward function, together with the transition function, represents a model of the environment. In model-based RL, the agent has access to an approximation of the transition function, either through experience or historical data collected by other agents, for instance. In this case, the RL problem is the MDP, and the solution is an optimal policy π, which can be determined using dynamic programming or policy evaluation, for instance.

Dynamic programming includes a set of algorithms that consist of breaking complex problems into sub-problems and combining their solutions. However, these algorithms have high requirements in terms of memory. Policy evaluation [25] is based on the iterative computation of the state-value function for a given policy. It is also known as the prediction problem. However, it becomes impractical in high-dimensional state and action spaces. Vehicular and cellular environments are highly stochastic and hard to formulate as MDPs. Hence, model-free algorithms are more popular and are adopted in such scenarios.

Model-Free RL: In model-free RL, a model of the environment is not required to find the optimal policy, as a model-free algorithm can estimate a value function or the policy directly through experience (i.e., interaction with the environment).

Value-Based RL Algorithms: Value-based RL algorithms are based on estimating the value, referred to as the expected return, of a given state at a time step. A value function enables the agent to determine how good it is to be in a state at a given time or to take an action from a state at a given time. For this reason, value functions are twofold: a state-value function and action-value function. Temporal difference (TD) [26] learning and its extension Q-learning [27] are state and action value-based algorithms, respectively. TD utilizes ideas from both Monte Carlo and dynamic programming. In fact, TD techniques use samples from the environment similarly to Monte Carlo methods, and, in the same way as dynamic programming methods, they use current estimates for updates through bootstrapping. The difference between Monte Carlo methods and TD is that while Monte Carlo methods only adjust their estimates once the final outcome is known, TD methods adjust predictions before the final outcome is known using later and better predictions about the future. This makes TD faster but less stable and might lead to wrong solutions.

In Q-learning, a function Q is used to map a state-action pair to the expected cumulative return (i.e., Q-value), allowing an agent to determine optimal actions in different states. The agent records all the actions maximizing the Q-value and thus infers an optimal policy defined by a list of optimal state-action pairs. However, in large state-action spaces, such as video games and robotics applications, calculating the Q-values of all the possible state-action pairs becomes impractical. Instead, the Q-value is approximated using deep neural networks (DNNs), for instance, in deep

Q-learning. Indeed, deep Q-networks (DQNs) [28] have surged in popularity and are widely adopted in many applications. Moreover, several other DNN-based models were proposed for deep Q-learning, including double-DQN and dueling DQN [29].

Furthermore, Q-learning is off policy. This means that the learning agent learns the value function using the action derived from another policy. SARSA (state action reward state action) [30] is a slight variation of technique. It is an on-policy and uses the action performed by the current policy to learn the Q-value. SARSA is typically preferable in situations where the agent's performance during the process of learning is important and failure is costly. For instance, in the case where the agent is an autonomous vehicle, it will be costly in terms of equipment if the vehicle crashes into an obstacle or falls down a cliff. In the learning process using Q-learning, since the agent always explores, there are more chances that the agent will fall off the cliff, for example.

Policy-Based RL Algorithms: Policy-based RL aims to learn a stochastic policy function that maps states to actions. While value-based RL algorithms optimize value function first, then derive optimal policies, the policy-based methods directly optimize an objective function (e.g., cumulative rewards). Examples of policy-based RL algorithms include REINFORCE. REINFORCE [31], also called Monte Carlo policy gradient, estimates the return by Monte Carlo methods using random episode samples and uses this estimate to update the policy parameter. The random sampling introduces high variability in the policy distribution and cumulative reward values, which leads to noisy gradients and unstable learning.

The policy-based approaches are highly suitable for very large or infinite action spaces. However, policy-based RL alone might be inefficient due to high variance and convergence to local optima. This can be enhanced through the usage of a baseline with REINFORCE. This idea is further explored in actor-critic methods.

Actor-Critic Algorithms: Actor-critic methods aim to combine strong points from both value-based and policy-based RL algorithms. To achieve this, the actor-critic methods build on TD methods, with the addition of a memory structure to represent the policy. The two components of actor-critic are an actor and a critic. The actor chooses which action to take based on a policy gradient approach, and the critic evaluates how good the action is and how it should adjust by computing the value function [32].

A key concept used in actor-critic methods is advantage, which refers to the difference between the actual outcome and the expected outcome. The advantage actor-critic–based methods are asynchronous advantage actor critic (A3C) [33] and advantage actor critic (A2C) [34]. A3C uses multiple workers in parallel environments, independently updating a global value function in an asynchronous manner. It was argued that the main benefit of having asynchronous actors is the efficient exploration of the state space. A2C is similar to A3C, but it only uses a single worker. Several experiments have found that A2C has comparable performance to A3C while being more resource efficient. Another actor-critic–based method is the deep deterministic policy gradient (DDPG) [35] algorithm, which has surged in popularity, as it can operate over continuous action spaces.

2.2.2.3 Artificial Neural Networks

ANNs are a set of ML models inspired by the brain. An ANN is composed of several sets of connected units called artificial neurons, which aim to mimic the neurons of a biological brain by learning to transmit signals to other neurons upon receiving a certain input. Each neuron is characterized by its parameters, namely weights and biases. An ANN has multiple successive layers of neurons, where the first layer is the input layer, then one or multiple hidden layers, followed by the output layer. Deep learning is a subfield of ML that encompasses many improvements in ANNs, namely deep neural networks, a special case of ANNs. A DNN refers to an ANN with multiple hidden layers. Many structures of ANNs were developed and serve different purposes. Examples of ANNs used in the CR-IoV include convolutional neural networks (CNNs) [36], which are widely used for image processing, and recurrent neural networks (RNNs) [37], which are suitable for sequential data.

2.2.3 Distributed Training and Inference of ML Models

The scarcity of historical training data, as well as the limited resources of edge devices, require leveraging collaborative and distributed algorithms for training and inference. Collaborative training can be deployed in a peer-to-peer fashion through gossip training or with the coordination of a centralized entity using federated learning. Inference, on the other hand, can make use of different parts or versions of models deployed on devices and on edge equipment.

2.2.3.1 Distributed Training

Gossip Training: Gossip training is a decentralized training method that aims to reduce training latency by using collaboration. It is built on randomized gossip algorithms, which refers to procedures of peer-to-peer communication that are based on the way epidemics spread. Such communications procedures have been adapted for ML. For instance, gossip averaging [38] can fast-converge towards a consensus among nodes by exchanging information peer to peer. Another example is GoSGD (gossip stochastic gradient descent) [39], which manages a group of independent nodes, where each of them trains a model following two steps iteratively: (1) each node updates its hosted model locally in a gradient update step, and (2) each node shares its information with another randomly selected node in a mixing update step. The steps are repeated until all the models converge on a consensus. The gossip distributed algorithms are fully asynchronous and decentralized, as they do not require any centralized nodes or variables. However, the fully decentralized and asynchronous aspects might not always be suitable for handling outliers and malicious nodes and statistical heterogeneity of the datasets' distributions.

Federated Learning: FL is a key technique for training ML models in a privacy-preserving fashion. Similarly to gossip training, FL is based on exchanging model updates instead of raw data. However, while gossip training is based on peer-to-peer communication, FL requires the orchestration of a centralized entity. Nonetheless, communication is the bottleneck of FL. Accordingly, several algorithms were developed to optimize FL in wireless and vehicular networks [40]. FL was also used as a key technique to enhance the performance of CR [41] and protect privacy in IoV applications [42].

2.2.3.2 Distributed Inference

DNN partitioning can be considered a fractional offloading approach that takes advantage of the multi-layer structure of DNNs. In this approach, some layers are computed on-device, and other layers are computed by the edge server or the cloud. This approach is based on the idea that after the first few layers of the DNN model have been computed, the size of the intermediate results is small compared to the original raw data, which makes them faster to send over the network to an edge server. Processing the data by passing it through some of the DNN layers can also add more privacy compared to sending raw data [43, 44].

2.3 ML CHALLENGES AND ENABLERS IN THE CR-IoV

So far, we have discussed ML as a subset of techniques for creating systems that can learn useful experiences and knowledge necessary for prediction and decision. However, the introduction of ML into CR in general and into the CR-IoV in particular raises a set of challenges related to the nature of the use case. Subsequently, a set of directives is also required to smoothly introduce ML into such a field. In this section, we provide a number of challenges one might face when introducing ML into the CR-IoV. We first discuss the integration of ML into CR, then ML into the IoV, and, finally, guidelines towards an ML-enabled CR-IoV. Figure 2.2 summarizes the set of considerations to enable ML in the CR-IoV.

Before digging further in the ML and its integration with CR and the IoV, a brief description of CR components is required, as well as for the IoV within the context of CR.

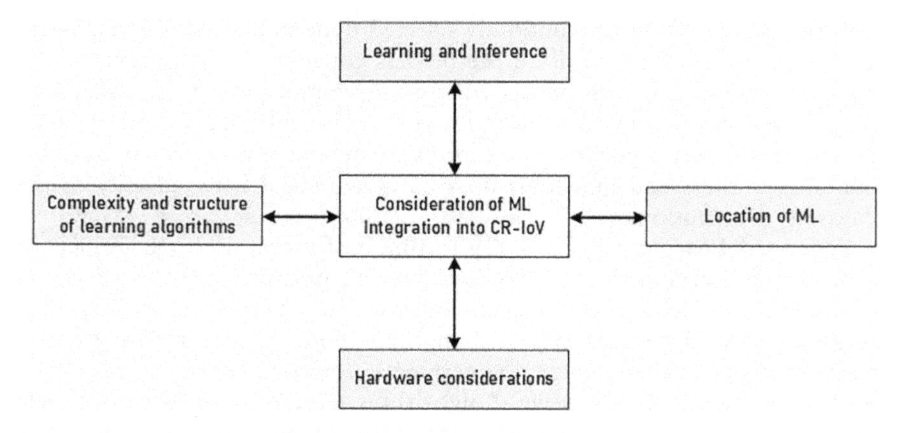

Figure 2.2 ML consideration for enabling the CR-IoV.

2.3.1 Definitions

2.3.1.1 Spectrum Sensing

Simply defined as the operation of acquiring spectrum-related information through attentive observation of radio resource frequency at a time and location of the surrounding environment. The major task of spectrum sensing is channel selection and identifying the primary vacant spectrum. Spectrum sensing provides the potential to detect spectral voids (in terms of frequency band, location, and time of availability) as well as primary user (PU) activity by periodically sensing the spectrum and using it free from any interference with other PUs.

Spectrum sensing is performed according to the following main approaches:

- **Cooperative Sensing** (CSS) provides a solution to improve sensing performance, where secondary users cooperate with one another to sense the spectral state of the channel (idle or busy). CSS makes the spectrum sensing process more reliable and accurate than individual CSS. It includes sharing local sensing results from multiple secondary users (SUs) to one central entity. CSS can be performed in a centralized manner, where the substrate of the network assumes the existence of a central entity that makes the decision based on the collected information from the SU [45]. CSS can also be performed in a decentralized fashion relying on full interaction between users requiring no central entity to make decisions [46]. Often solutions come with a consensus-based solution that puts no condition on knowing the network topology. However, CSS requires supplementary overhead exchanges which in return affect the energy consumption, thus causing additional sensing delay. Moreover, CSS-acquired information swiftly gets deprecated, especially in environments with high mobility.

- **Non-Cooperative Sensing** (non-CSS) sensing, or PU transmitter detection, is performed individually by each SU to sense the presence of a PU's activity [47]. Unlike CSS, the detection methods adopted in non-CSS require small overheads; however, it depends on the availability of the network architecture. The infrastructure might not always be available and could be subject to noise, interference, and sometimes hidden PUs. An example of non-CSS detection is energy-based detection consisting of measuring the energy levels of the received signal over a spectrum band. The results of ED are then compared to a carefully chosen threshold. The presence or absence of a PU is obtained based on comparing the results with the threshold. Another method for non-CSS detection is wavelet-based detection. This method leverages the fast Fourier transform and wavelet theory to detect a PU signal through detecting irregularities in the power spectral density. Matched filter-based detection uses linear filters to maximize the output signal to noise ratio (SNR) of a signal to detect a PU. Other methods for non-CSS in the literature are covariance matrix-based detection and cyclostationary feature-based detection.
- The ***Hybrid Sensing*** technique allows SUs to share their sensing data in a decentralized fashion to identify vacant channels [48]. In the presence of a PU, the channel is immediately evacuated with a prior alert. The detection time in such an approach is significantly reduced; however, to achieve improved performance, dedicated equipment is required to ensure cooperation [49].

In the context of this chapter, ML learning in the CR-IoV can be adopted in the following:

- Mobility issues, especially high-speed scenarios
- Spectrum sensing synchronization
- Delay and latency improvement
- Improving adaptation to a dynamic environment
- Security of CR-vehicular networks

2.3.1.2 Mobility Management

The main idea of CR networks is to make unlicensed devices utilize a licensed spectrum with no interference with primary users. When a PU appears or reappears on a channel, all SUs should move to another vacant channel to escape an interference situation. Such an approach is also referred to as spectrum mobility [50]. Spectrum mobility comprises two important processes, spectrum handoff and connection management. Many handoff strategies exist, for instance, non-handoff, pure reactive handoff, pure proactive handoff, and hybrid handoff [51]. In addition, mobility management includes location management and handoff management. ML would play an important role to improve the CR-IoV from the perspective of mobility management through providing efficiency to mobility management processes.

2.3.1.3 Spectrum Sharing

Spectrum sharing is the process of administering the allocation of spectrum among CR users (vehicles in the case of the IoV) while at the same time maintaining the network's overall performance. There are several perspectives of spectrum sharing, including based on utilization and based on the adopted sharing paradigm (i.e., distributed and centralized). From a spectrum utilization perspective, there are unlicensed sharing and licensed sharing. In unlicensed mode, all users have the same priority, while in licensed mode, PUs have a higher priority. SUs can only access both types of spectrum sharing when PUs are not present. Concerning the sharing paradigm adopted, the case of centralized spectrum sharing requires a central node to control spectrum allocation. The distributed spectrum sharing case gives individual nodes control over spectrum allocation. Other types of spectrum sharing in CR are cooperative and non-cooperative sharing. It is worth mentioning here that spectrum sharing can be performed based on the access technology as well. Opportunistic spectrum access, where the co-existence of PUs and SUs is not allowed, requires SUs to opportunistically find empty holes in channel and occupy them. Underlay spectrum sharing is where PUs and SUs are allowed to co-exist on the condition that SUs not cause interference with PUs (i.e., SU signal power stays under a given previously configured threshold). SUs in this case adopt a spread spectrum approach to maintain lower signal power. Finally, in overlay spectrum sharing, SUs are allowed to simultaneously coexist with PUs on the same channel under the condition of cooperative transmission, where SUs help PUs in transmitting.

2.3.2 The CR-IoV

The IoV is characterized by a number of distinctive features, such as a dynamic network substrate, scalability, reliability, resourceful capabilities, and high integration with other users' devices. The IoV has dominated transportation systems due to numerous special traits, like dynamic topological structures, huge network scale, reliable internet connection, compatibility with personal devices, and high processing capability. The IoV can take advantage of CR-related features for communication purposes. The IoV faces several issues and challenges, especially when it comes to transmitting heavy amounts of data through the wireless medium. Data amounts represent a challenging parameter in enabling the IoV's full potential in managing road traffic, safety, and infotainment services. Resource allocation approaches in the IoV are ineffective to overcome transmission challenges, which already suffer spectrum scarcity issues. CR can help the IoV overcome challenges related to spectrum scarcity problems. Enabling CR in the IoV gives birth to the CR-IoV paradigm. The CR-IoV is a promising paradigm where CR technologies such as dynamic spectrum access and cooperative communications will increase the potential of the IoV in achieving efficient delivery of vehicular services.

2.3.3 ML in CR

CRs are environment-aware wireless agents capable of learning from their environment to adapt to different situations. An example of this situation is to switch channels when detecting the presence of PUs. CR must acquire the ability to gather experience from past and current observations, in addition to operating within different spectrum channels. CRs have different levels of channel state information, varying over the level of available information, which is sometimes complete, incomplete, or totally unknown information. The difference in spectrum parameter values, namely wireless channel wavering, error quantization, and estimation, makes the process of acquiring channel information unreliable. ML therefore can provide CR with the needed intelligence to estimate actions under the existence of other CRs and consequently efficiently manage the spectrum. ML integration into CR takes place at several levels of the CR process. For instance, ML will boost the spectrum sensing process with intelligence to adapt to a dynamic network environment, network overhead and latency, sensing information synchronization, and different QoS requirements. ML also will provide efficient prediction of PU behavior, as well as spectrum handoffs and optimal routing in a CR environment.

2.3.4 ML in the IoV

The highly mobile environment of the IoV will profit from ML advantages to handle constraints imposed by the use case. Road safety scenarios can be improved using ML. For instance, helping drivers to increase their vigilance, assisting lane changing, traffic sign recognition, and other existing driver assistance features. Traffic congestion management using ML consists of predicting traffic flow loads and therefore providing alternative solutions such as rerouting. ML is also applied to smart parking, load balancing, toll systems, and traffic signaling. Resource allocation/provisioning for both energy and spectral efficiency will become more efficient. Finally, infotainment will also benefit from ML in scenarios of entertainment, vehicular information exchange, advertisement, and scheduling.

2.3.5 Towards a ML-Enabled CR-IoV

In this subsection, we provide different ways in which ML can be integrated into the CR-IoV. The integration differs from one ML paradigm to another.

In the previous subsections, we provided manners in which ML was leveraged in the CR and IoV independently.

The integration of ML into the CR-IoV takes many shapes and levels. Indeed, ML is a powerful tool that will drastically improve CR-IoV performance through different approaches (e.g., reinforcement learning, supervised/unsupervised learning, deep learning). However, executing ML algorithms needs to be carefully designed due to the complicated and heavy structures that require significant amounts of computational resources. Therefore, it is important to optimize the learning process on many levels.

Computation as the cornerstone of ML has also known a giant leap in terms of development. Several works proposed dedicated hardware to better enhance the time and energy of ML methods. Such an interest in hardware development got attention thanks to the deployments of DNNs on graphics processing units (GPUs). Hardware used in ML is generally categorized into general-purpose hardware and dedicated-purpose hardware. General-purpose hardware such as advanced reduced instruction set computer (RISC) machines (ARM)-based computation or GPU-based computation (e.g., Raspberry Pi and Nvidia Jetson) are widely used for embedded applications. Dedicated-purpose hardware often refers to recently emerged tensor processing units (TPUs) such as Intel's Movidius and Google's Coral.

- The complexity of the algorithms means that the adopted/adapted ML approach should take into consideration the complexity of the problem being tackled with ML. At this stage, the parameters and their values need to be carefully chosen [**52**], for instance, the number of layers in a NN, the learning rate, and the loss function. Methods like principal component analysis are used to extract important features to consider when targeting better learning performance.
- The nature of the algorithms denotes if the algorithm being executed is in learning mode or inference mode. Resource-wise, the learning process requires huge amounts of computation in order to achieve adequate convergence. On the other hand, the inference of already trained models also has an important impact on the performance of the system. Recent advances in ML propose lightweight designs for different ML algorithms and approaches. It is worth mentioning CNNs, RNNs, spiking neural networks (SNNs), SVMs, and decision trees here. Other methods dealing with the same challenges include adopting transfer learning approaches and knowledge distillation.
- Location and resources utilized refer to the facility where the execution is being performed (at the user side or the service provider side) and the kind of hardware is being used. On the user side, the learning process should take into consideration the resource constraints from an energy and computation perspective. Energy-efficient ML techniques have been widely investigated; however, the focus is always on inference. ML algorithms, specifically the training, take two forms (offline

and online). In offline modes, the training takes place on dedicated hardware, and only the results are executed online. An example of infrastructure allowing such a paradigm is Open-RAN (O-RAN) [53, 54]. In such a paradigm, training is based on previously collected datasets (spectrum sensing from previous rounds, mobility-related data). As for online algorithms, the training and execution take place at the same location within the same moment of data collection.

2.4 ML APPLICATIONS IN THE CR-IoV

ML applications in a CR-IoV environment are growing considerably, with the aim of bringing more automation to the operation of entities in an IoV network. This automation enables the development of models capable of making relevant and timely decisions, which improves the QoS provided to IoV users. ML applications in CR-IoV include: (1) spectrum sensing and channel allocation, (2) road safety, (3) traffic congestion, (4) security issues, and (5) infotainment.

2.4.1 Spectrum Sensing and Channel Allocation

In CR networks, secondary users, that is, unlicensed users, can access the licensed spectrum resources of primary users in an opportunistic manner that avoids interference between them. Spectrum sensing is a key function of CR technology that enables secondary users to sense channels and identify free frequency bands, called "spectrum holes" or "white spaces". Accordingly, spectrum sensing must be highly reliable to avoid and mitigate interference between primary and secondary users in order to promote the utilization of spectrum resources.

Spectrum sensing in the CR-IoV is a challenging task due to the dynamic nature of the IoV environment, the volatile activity of primary users, and the heterogeneous QoS requirements of users. Indeed, the high mobility of vehicles and the rapid variation in network density impact the spectrum sensing operation, which must consider the trade-off between accurate channel sensing and fast channel sensing to identify unoccupied resources. Primary user activity, such as transmission time on channels, is very difficult to profile in a dynamic environment like that of the IoV. The QoS requirements of users in an IoV environment depend on the required applications, such as safety and non-safety applications. Therefore, ensuring access to resources that meet the QoS requirements of each user by the spectrum sensing operation is increasingly complicated. To overcome all these challenges, several ML techniques can be applied.

In an IoV environment, vehicles, as secondary users, must accurately identify spectrum holes in a short period of time in order to leverage them before moving to another area where they may not be available. In addition, the abrupt and potentially unexpected fluctuations in IoV network traffic due to rapid changes in network topology impose more constraints to spectrum sensing operations. ML provides more reliable and faster spectrum sensing in such a dynamic environment. ML algorithms can deal with rapid changes in the CR-IoV environment by learning from data related to the environment characteristics, primary user features, and secondary user sensing parameters, such as signal-to-noise ratio and location. Then, they predict spectrum holes and make decisions to improve spectrum utilization. In particular, RL algorithms are able to learn from several input parameters related to the IoV environment and the activity of the primary users. In [55], signal-to-noise ratio, channel quality, vehicle mobility, and vehicle density are used to achieve better spectrum sensing efficiency and to allocate available channels that meet the QoS required by the vehicles.

To overcome issues related to primary user activity, RL algorithms are an efficient solution. A RL algorithm can learn the activity of the primary user's traffic pattern to forecast which channel will potentially be available for use by secondary users in the future during congestion. For instance, a RL agent, which can be embedded in a roadside unit (RSU), can interact with the IoV environment during congestion to guide vehicles to sense particular channels [56]. For this purpose, the RL agent needs to learn which channels have a high throughput and remain inactive for long periods. To train a model that can identify such channels, the RL agent takes an action by selecting the channels to send to vehicles. Then, vehicles perform spectrum sensing operation and report to the RL agent the channel state. Based on the spectrum sensing results, the RL agent associates a reward with each channel. This reward is related to the availability of the channel, the bandwidth of the channel, and how long the channel has been leveraged by a secondary user. The higher the reward of a channel, the more this channel will be selected by the RL agent to be sent to vehicles for spectrum sensing operations. To improve the learning of free channel detection, the probability of missed detection or wrong alarms about the presence of primary users, which can significantly reduce the efficiency of spectrum detection operations, can be taken into account in the reward function. In other words, considering the spectrum sensing error in the learning procedure increases the probability of taking good actions. For instance, [57] proposes a Q-learning algorithm that will be used by vehicles to select channels for IoV communications. This algorithm penalizes actions resulting from erroneous detection of primary users' presence. In addition, the reward function can also depend on the bandwidth of the channel selected by a secondary user as well as the transmission power selected by a secondary user when a primary user is present in the selected channel. Indeed, if the selected transmit power is high, it may lead to significant interference for the primary user. Thus, taking such actions should be penalized. On the other hand, the greater the bandwidth of the selected channel, the better the action chosen by a secondary user and hence the higher the impact on the reward function.

Deep reinforcement learning (DRL) algorithms are able to provide a compact low-dimensional representation from high-dimensional data. Accordingly, they can learn from several parameters as input data to select the channels to assign to secondary users in an IoV network. To identify channels with the minimum probability of primary user arrival, channel occupancy, channel occupancy consistency, and channel capacity are used in [58] to train DRL models. The input data are collected by RSUs, which sense the channels continuously and collect data on channel occupancy by primary users every second of a given minute on a particular day. The collected data are shared between RSUs that constitute an RSU cluster. Then, each RSU uses these data to train its DRL model. The trained models are used to select channels with the minimum probability of primary user arrival that will be given to vehicles. To design an optimal data transmission scheduling scheme in a CR-IoV network, the communication mode, that is, V2V or V2I; the caching capability of vehicles; and the delay constraint are considered in the DRL model training phase [59]. In fact, the action of selecting a channel to provide data to the network infrastructure is related to the chosen communication mode. The learning process of the proposed DRL model depends on the correlation between the communication model, the QoS requirement in terms of delay, and the size of the data queued in the vehicles' cache. A control center collects information on vehicle states, that is, vehicle cache status and mobility. Then, it schedules data transmission for each vehicle on a specific channel through a communication mode.

2.4.2 Road Safety

The IoV network provides effective road safety applications that ensure the safety of road users. These applications involve monitoring traffic and the road, ensuring the communication of safety messages between vehicles, and collecting or sharing information about the road surface and curves. CR technology has enhanced IoV road safety applications by providing dynamic access to spectrum and prioritizing access to radio resources for these applications. However, road safety applications require accurate situational awareness to improve the comfort and convenience of drivers and passengers. For this reason, vehicles and road infrastructure are equipped with various sensors to collect timely information about what is happening around the vehicles. The data collected by these sensors are huge, which provides an opportunity to conduct accurate and robust safety analysis. To incorporate such features in the CR-IoV network, ML algorithms enable efficient analysis to predict unexpected situations that might occur in the CR-IoV network and which can impact safety applications.

V2X safety applications, such as collision prevention at urban intersections, emergency electronic brake light warnings, lane change and highway passing warnings, and roadwork warnings, may require road perception and control information from the road infrastructure and all road users via V2X communications. To support these applications, ultra-reliable low-latency communications (URLLC) must be supported by the CR-IoV network. According to 3GPP specifications, URLLC communications require 99.9% reliability for a single frame of data and an end-to-end latency of a few milliseconds. To meet these requirements, road safety data should be prioritized by allowing rapid access to the spectrum. Indeed, the channels with the best conditions can be reserved for safety application data and the remaining channels for non-critical data, such as infotainment applications. This can be achieved through ML algorithms. In **[60]**, channel state information, user data type (i.e., critical and non-critical data), and number of users are used to train a ML algorithm. The proposed ML algorithm can be implemented in base stations close to roads and attached to a multi-access edge computing (MEC) server, which can collect data from vehicles to train the ML model.

To mitigate interference with primary users and to meet the QoS requirements of safety applications, ML algorithms can use historical spectrum sensing data to predict and allocate the available spectrum holes to secondary users. In fact, prior knowledge about spectrum holes is leveraged to capture spatial correlation among spectrum sensing data gathered from vehicles that travel along the same road. In particular **[61]**, the channel steady-state probability, channel state transition probability, and hidden channel state are utilized to predict the occurrence probability of spectrum holes and assign them to vehicles to transmit messages of security applications.

2.4.3 Security Issues

Security in the IoV network is a serious issue, since any failure in the system directly impacts the safety of the users. Although CR improves communications in the IoV environment, it raises several security concerns. Indeed, the CR system is vulnerable to security threats due to its intelligence functionality, the heterogeneity of wireless access technologies and system/network operators, and the dynamic application of spectrum access. Accordingly, the same security issues arise in the CR-IoV environment. These security problems result from failure to meet security requirements, namely confidentiality, integrity, access control, or authentication. To address these issues, ML algorithms are leveraged to detect attacks and handle various security threats in CR-IoV communications. ML techniques can predict attacks more quickly and accurately, helping to protect CR-IoV communications from security threats that degrade data confidentiality, integrity, authentication, and availability. This is due to their ability to examine traffic data, identify the type of attack, and then distinguish between malicious and benign CR-IoV network nodes.

Cooperative spectrum sensing enables reliable detection of spectrum holes, since it exploits the spatial diversity of secondary users, that is, vehicles, who share their sensing measurements to make joint decisions. However, this collaborative sensing approach increases CR-IoV vulnerability to some potential attacks, such as data falsification attacks. In this attack, a malicious secondary user falsifies its spectrum sensing report, which will be disseminated to other secondary users. As a result, spectrum sensing performance deteriorates, which can either lead to interference with primary users or ruin radio resource access opportunities for secondary users. To detect such anomalous behavior of a secondary user, several ML classifiers, namely ANN, SVM, RF, and KNN [**62, 63**], can be used. Such algorithms use historical sensing reports to learn the footprint of data-falsifying attackers. Then, they can effectively classify secondary users as benign or malicious by analyzing their spectrum sensing report. In certain scenarios, data-falsifying attackers dynamically adapt their behavior to avoid the features and patterns learned by the classifiers. Therefore, more advanced ML-based defense approaches must be employed to identify any potential attackers. The adversarial machine learning technique can be used to generate data falsification attacks and thus to train models on what an adversarial attack could possibly look like [**64**]. Data falsification can involve the position of vehicles in a CR-IoV network. In [**13**], some features related to vehicle positions, such as the angle of arrival, the distance between the transmitter and receiver, and the difference between the estimated distance and the reported distance are used to detect falsification of position data. In detail, an ensemble learning approach is proposed to design an intrusion detection system that recognizes misbehaving vehicles. The ensemble learning approach combines a KNN algorithm and a RF algorithm to detect position falsification attacks. It consists of a combination of classifiers that perform learning in a central entity of the IoV network, and then the resulting intrusion detection system is shared with vehicles to cooperatively participate in the detection of data-falsifying attackers.

The jamming attack is one of the major threats to spectrum-sharing wireless networks, especially for CR networks. In such an attack on a CR network, a malicious attacker transmits radio signals in order to disrupt the existing communications of secondary users. These signals inject interference in the communications, which decreases the signal-to-noise ratio on the receiver side. When a CR-IoV network suffers from a jamming attack, the wireless signals, for example, from vehicles, are overwhelmed by inappropriate radio interference signals, which makes it difficult for the IoV network devices to decode the received data. Hence, a failure to receive data can lead to serious dangers, especially if the transmitted data contain important information, such as data from traffic safety applications. ML-based solutions can provide an optimal defense against jamming attacks in the CR-IoV networks. For instance, FL is a powerful ML technique for detecting jamming attacks. Indeed, since in FL the data are trained locally, for example, in CR-IoV devices, and then sent to a central entity to perform model aggregation, FL can extract fine-grained properties of jamming data [**65**]. In other words, based on FL techniques, the detection of jamming attacks will be performed at the device level. Furthermore, FL allows training a global model using local models obtained from diverse and heterogeneous datasets. Then, the local models are permanently updated using the global model, making them more efficient in performing on-device jamming attack detection. Accordingly, FL not only provides a high-performance jamming attack detection system but also protects the central entity, such as RSUs. RL techniques can also be used to secure CR-IoV networks against jamming attacks without knowing the jamming patterns related to vehicular communications. In [**66**], an RL algorithm is proposed to prevent a jammer from predicting the relay strategy employed by unmanned aerial vehicles (UAVs) to relay messages from sensors to the RSUs. In fact, a smart jammer observes communications between sensors, for example, onboard units, and RSUs and chooses an appropriate jamming power to disrupt this vehicular communication. Using a RL algorithm, a UAV can decide whether to relay a sensor message to a distant RSU. Therefore, it is very difficult for a jammer to carry out its attack, since it cannot forecast the relaying strategy used by the UAVs.

2.5 CHALLENGES, ISSUES, AND FUTURE DIRECTIONS

While current efforts in favor of the effective use of ML in the CR-IoV are ongoing, the road towards full integration of both technologies is still long and poses many challenges. In addition to challenges purely related to designing CR for the IoV, enabling ML for the CR-IoV poses further additional challenges. These challenges include state and network dynamics, multi-agent RL in the CR-IoV, distributed learning, scalability and universality, and proactive learning. Figure 2.3 presents the challenges in the ML-based CR-IoV.

2.5.1 Machine Learning–Related Challenges

Choosing the right ML approach to solve CR-IoV–related problems is quite a challenging task. The adopted ML approach should be suitable to the nature of the environment for which ML is designated. The CR environment is a highly dynamic environment which changes rapidly over time. Mostly, adopting any ML approach comes with a set of advantages and issues. By way of illustration, supervised NNs rely on data labeling to provide abstracted classification and require training under different conditions imposed by the environment. The results of training depend strongly on the initial parameters. Consequently, data quality and selection should be attentively tackled. Another common challenge among ML approaches is the convergence time, which depends on the network size. Increasing the size of the network often leads to slower convergence time. In addition, increasing the number of hidden layers may offer improved efficiency; nevertheless, with a significantly large size of training data, the training process becomes time consuming.

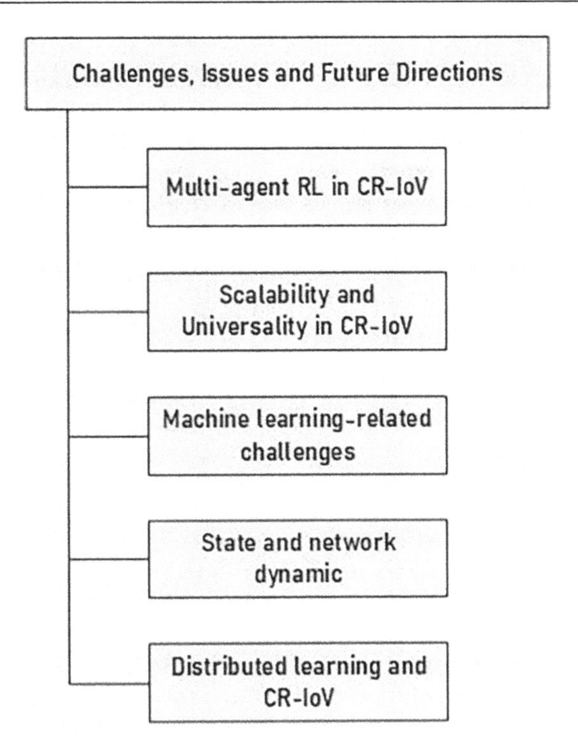

Figure 2.3 Summary of challenges in CR-IoV.

2.5.2 State and Network Dynamics

RL in general and DRL in particular is widely utilized in different CR-related use cases on the one hand. On the other hand, the IoV imposes constraints related mainly to high mobility and safety scenarios. Applying DRL provides adequate policies regarding an objective function (e.g., to enhance the prediction of PU arrival for an improved handoff operation) without having complete/precise network state information. For such a reason, users are usually compelled to provide their local state. The local state is related to the observed environment, which can highly increase the computation time when interacting with a sufficiently high number of users. Additionally, the adopted RL approach plays a crucial role in dealing with the evolution and dynamics of the system; hence the complexity of the observations.

2.5.3 Multi-Agent RL (MARL) in the CR-IoV

By definition, MARL is a learning approach gathering autonomous agents (e.g., vehicles on a highway) sharing the same environment. The learning process in MARL considers the interaction of agents with the environment, and with other agents, which makes it difficult. Q-learning considers the agents part of the environment, which makes the convergence process significantly unstable. Additionally, changing the policy of an agent will impact the policies of the other agents. From a single learning perspective approach, conjecturing policies of other agents would be used as solution to overcome the changing of policy. Another approach to overcome such a challenge is studying agent interactions using game theory, specifically the mean-field games that are more adapted to these cases [67, 68].

2.5.4 Distributed Learning and the CR-IoV

The CR-IoV requires both communication and computation to provide the effectiveness and efficiency of deep learning comes from using algorithms with highly complex structure however IoV is resource limited infrastructure. As a solution, MEC infrastructure can be leveraged [69] to provide sufficient resources for ML purposes. Another challenge related to distributed learning is the hardware used. Distributed learning requires advanced dedicated hardware for the training process; a recent deployment suggested using GPUs and TPUs. Distributed leaning requires a significant amount of storage, which is an issue in the CR-IoV due to the limited resources of vehicles. An alternative solution comes in the form of storage area networks (SANs). Scalability is also worth investigating in addition to universality in the CR-IoV.

2.5.5 Scalability and Universality in the CR-IoV

The main idea behind ML is to dispose of human intervention through making self-learning systems. The targeted system should be able to adapt to different possible scenarios and use cases. Nevertheless, most literature on applying ML in the CR-IoV considers a set of assumptions specific to an optimization problem for a given objective function. For instance, models being applied to an urban scenario cannot be adopted for a rural environment. Such a constraint is imposed by the structure of the adopted ML approach (e.g., CNN applied to mobility). To cope with such an issue, transfer learning can be adopted in dynamic routing for the IoV, which transfers the trained neural structure or a part of it to other entities. This helps reduce/avoid repetitive training, thus reducing resource utilization. Although transfer learning is not scalable, it only makes sense when the structure is being transferred among related tasks. Universality can be reached by making the ML more intelligent. Newly emerged meta-learning is proposed to make the ML itself learn and self-improve through changing self-structures and self-parameters. Meta-learning, therefore, represents a very interesting research direction.

2.6 CONCLUSION

Cognitive radio networks are considered an innovative paradigm to expand the use of the spectrum through the utilization of unused/underused frequencies in wireless environments. The Internet of Vehicles constitutes the cornerstone for future intelligent transportation systems. The IoV is intended to improve the performance of vehicular ad hoc networks through their integration into Internet of Things architectures. In IoV scenarios, vast amounts of data are being exchanged over the wireless network, which requires the development of efficient spectrum allocation strategies. CR can improve the IoV through providing sets of tools allowing spectrum management, mobility and handoffs, security, and dynamic resource management. However, CR cannot be integrated/applied to the IoV due to the nature and features of the environment that change rapidly. ML consists of training a given system to adapt to different situations regarding decision-making in an autonomous fashion. Through ML, CR will be equipped with the required features to overcome its integration constraints in the IoV. This chapter represents an overview of ML approaches used for CR-IoV benefits. First, we gave a comprehensive overview of CR in the IoV. Subsequently, we provided various ML techniques and their challenges regarding the CR-IoV. Then, we presented different ML applications in the CR-IoV from literature, specifically spectrum sensing, mobility management, road safety, and security issues. Finally, we presented different challenges regarding ML integration into the CR-IoV. These challenges include full ML challenges such as the curse of dimensionality and convergence to a stable policy. Network dynamics are also a crucial feature to consider when designing ML-based solutions for the CR-IoV due to constraints of scalability.

REFERENCES

1. S. Sharma and B. Kaushik, "A survey on Internet of Vehicles: Applications, security issues & solutions," *Vehicular Communications*, vol. 20, p. 100182, 2019.
2. B. Wang and K. R. Liu, "Advances in cognitive radio networks: A survey," *IEEE Journal of Selected Topics in Signal Processing*, vol. 5, no. 1, pp. 5–23, 2010.
3. Y. Wang, X. Li, P. Wan, and R. Shao, "Intelligent dynamic spectrum access using deep reinforcement learning for VANETs," *IEEE Sensors Journal*, vol. 21, no. 14, pp. 15554–15563, 2021.
4. E. S. Ali, M. K. Hasan, R. Hassan, R. A. Saeed, M. B. Hassan, S. Islam, N. S. Nafi, and S. Bevinakoppa, "Machine learning technologies for secure vehicular communication in Internet of Vehicles: Recent advances and applications," *Security and Communication Networks*, vol. 2021, 2021.
5. F. Tang, B. Mao, N. Kato, and G. Gui, "Comprehensive survey on machine learning in vehicular network: Technology, applications and challenges," *IEEE Communications Surveys & Tutorials*, vol. 23, no. 3, pp. 2027–2057, 2021.

6. S. R. Pokhrel and S. Singh, "Compound TCP performance for Industry 4.0 WiFi: A cognitive federated learning approach," *IEEE Transactions on Industrial Informatics*, vol. 17, no. 3, pp. 2143–2151, 2020.
7. Y. Zhang, Q. Wu, and M. Shikh-Bahaei, "Vertical federated learning based privacy-preserving cooperative sensing in cognitive radio networks," in *2020 IEEE Globecom Workshops (GC Wkshps)*, pp. 1–6, IEEE, 2020.
8. F. A. Awin, Y. M. Alginahi, E. Abdel-Raheem, and K. Tepe, "Technical issues on cognitive radio-based Internet of Things systems: A survey," *IEEE access*, vol. 7, pp. 97887–97908, 2019.
9. W. S. H. M. W. Ahmad, N. A. M. Radzi, F. Samidi, A. Ismail, F. Abdullah, M. Z. Jamaludin, and M. Zakaria, "5G technology: Towards dynamic spectrum sharing using cognitive radio networks," *IEEE Access*, vol. 8, pp. 14460–14488, 2020.
10. N. S. Altman, "An introduction to kernel and nearest-neighbor nonparametric regression," *The American Statistician*, vol. 46, pp. 175–185, August 1992. Publisher: Taylor & Francis ePrint: www.tandfonline.com/doi/pdf/10.1080/00031305.1992.10475879.
11. J. R. Quinlan, "Learning decision tree classifiers," *ACM Computing Surveys*, vol. 28, pp. 71–72, March 1996.
12. L. Breiman, "Random forests," *Machine Learning*, vol. 45, pp. 5–32, October 2001.
13. R. Tibshirani, "Regression shrinkage and selection via the lasso," *Journal of the Royal Statistical Society. Series B (Methodological)*, vol. 58, no. 1, pp. 267–288, 1996. Publisher: Royal Statistical Society, Wiley.
14. D. E. Hilt and D. W. Seegrist, *United States and Pacific Northeastern Forest Experiment Station (Radnor, Ridge, a Computer Program for Calculating Ridge Regression Estimates)*. Upper Darby, PA: Department of Agriculture, Forest Service, Northeastern Forest Experiment Station, 1977, pp. 1–10.
15. C. Cortes and V. Vapnik, "Support-vector networks," *Machine Learning*, vol. 20, pp. 273–297, September 1995.
16. J. MacQueen, "Some methods for classification and analysis of multivariate observations," *Proceedings of the Fifth Berkeley Symposium on Mathematical Statistics and Probability, Volume 1: Statistics*, vol. 5.1, pp. 281–298, January 1967. Publisher: University of California Press.
17. O. Maimon and L. Rokach, *Data Mining and Knowledge Discovery Handbook*. NewYork: Springer Science & Business Media, May 2006.
18. J. Li, K. Cheng, S. Wang, F. Morstatter, R. P. Trevino, J. Tang, and H. Liu, "Feature selection: A data perspective," *ACM Computing Surveys*, vol. 50, pp. 1–94, December 2017.
19. I. Guyon, S. Gunn, M. Nikravesh, and L. A. Zadeh, *Feature Extraction: Foundations and Applications*. Springer, November 2008. Google-Books-ID: FOTzBwAAQBAJ.
20. F. J. Ferri, P. Pudil, M. Hatef, and J. Kittler, "Comparative study of techniques for large-scale feature selection," in *Machine Intelligence and Pattern Recognition, Vol. 16 of Pattern Recognition in Practice IV* (E. S. Gelsema and L. S. Kanal, eds.). New York, USA: North-Holland, January 1994, pp. 403–413.
21. E. Bingham and H. Mannila, "Random projection in dimensionality reduction: Applications to image and text data," in *Proceedings of the Seventh ACM SIGKDD International Conference on Knowledge Discovery and Data Mining,*

KDD '01, (New York, NY, USA), pp. 245–250, Association for Computing Machinery, August 2001.
22. A. Hyvärinen, "Independent component analysis: Recent advances," *Philosophical Transactions of the Royal Society A: Mathematical, Physical and Engineering Sciences*, vol. 371, p. 20110534, February 2013. Publisher: Royal Society.
23. J. Forkman, J. Josse, and H.-P. Piepho, "Hypothesis tests for principal component analysis when variables are standardized," *Journal of Agricultural, Biological and Environmental Statistics*, vol. 24, pp. 289–308, June 2019.
24. R. Bellman, "A Markovian decision process," *Journal of Mathematics and Mechanics*, vol. 6, no. 5, pp. 679–684, 1957. Publisher: Indiana University Mathematics Department.
25. C. Dann *et al.*, "Policy evaluation with temporal differences: A survey and comparison," *Journal of Machine Learning Research*, vol. 15, pp. 809–883, 2014.
26. G. Tesauro *et al.*, "Temporal difference learning and TD-gammon," *Communications of the ACM*, vol. 38, no. 3, pp. 58–68, 1995.
27. F. S. Melo, "Convergence of Q-learning: A simple proof," Institute of Systems and Robotics, Tech. Reports, pp. 1–4, 2001.
28. V. Mnih *et al.*, "Human-level control through deep reinforcement learning," *Nature*, vol. 518, no. 7540, pp. 529–533, 2015.
29. M. Sewak, "Deep Q network (DQN), double DQN, and dueling DQN: A step towards general Artificial Intelligence," in *Deep Reinforcement Learning: Frontiers of Artificial Intelligence*. Springer, 2019, pp. 95–108.
30. R. S. Sutton and A. G. Barto, "Reinforcement learning: An introduction," *Robotica*, vol. 17, no. 2, pp. 229–235, 1999.
31. R. S. Sutton, D. McAllester, S. Singh, and Y. Mansour, "Policy gradient methods for reinforcement learning with function approximation," *Advances in Neural Information Processing Systems*, vol. 12, 1999.
32. I. Grondman, L. Busoniu, G. A. D. Lopes, and R. Babuska, "A survey of actor-critic reinforcement learning: Standard and natural policy gradients," *IEEE Transactions on Systems, Man, and Cybernetics, Part C (Applications and Reviews)*, vol. 42, no. 6, pp. 1291–1307, 2012.
33. M. Babaeizadeh, I. Frosio, S. Tyree, J. Clemons, and J. Kautz, "Reinforcement learning through asynchronous advantage actor-critic on a GPU," *arXiv preprint arXiv:1611.06256*, 2016.
34. V. Mnih, A. P. Badia, M. Mirza, A. Graves, T. P. Lillicrap, T. Harley, D. Silver, and K. Kavukcuoglu, "Asynchronous methods for deep reinforcement learning," *Proceedings of the 33rd International Conference on Machine Learning*, New York, NY, USA, 2016. JMLR: W&CP volume 48.
35. T. P. Lillicrap, J. J. Hunt, A. Pritzel, N. Heess, T. Erez, Y. Tassa, D. Silver, and D. Wierstra, "Continuous control with deep reinforcement learning," *arXiv:1509.02971 [cs, stat]*, July 2019. arXiv: 1509.02971.
36. S. Albawi, T. A. Mohammed, and S. Al-Zawi, "Understanding of a convolutional neural network," in *2017 International Conference on Engineering and Technology (ICET)*, pp. 1–6, August 2017.
37. H. Salehinejad, S. Sankar, J. Barfett, E. Colak, and S. Valaee, "Recent advances in recurrent neural networks," *arXiv:1801.01078 [cs]*, February 2018. arXiv: 1801.01078.
38. A. Ghosh, B. Prabhakar, and D. Shah, "Randomized gossip algorithms," *IEEE Transactions on Information Theory*, vol. 52, no. 6, pp. 2508–2530, 2006.

39. M. Blot, D. Picard, M. Cord, and N. Thome, "Gossip training for deep learning," *arXiv:1611.09726 [cs, stat]*, November 2016. arXiv: 1611.09726.
40. A. Taïk, Z. Mlika, and S. Cherkaoui, "Clustered vehicular federated learning: Process and optimization," *IEEE Transactions on Intelligent Transportation Systems*, pp. 1–13, 2022.
41. S. R. Pokhrel and S. Singh, "Compound TCP performance for Industry 4.0 WiFi: A cognitive federated learning approach," *IEEE Transactions on Industrial Informatics*, vol. 17, pp. 2143–2151, March 2021. Conference Name: IEEE Transactions on Industrial Informatics.
42. Z. Yu, J. Hu, G. Min, Z. Zhao, W. Miao, and M. S. Hossain, "Mobility-aware proactive edge caching for connected vehicles using federated learning," *IEEE Transactions on Intelligent Transportation Systems*, vol. 22, no. 8, pp. 5341–5351, 2021.
43. H. Li, K. Ota, and M. Dong, "Learning IoT in edge: Deep learning for the Internet of Things with edge computing," *IEEE Network*, vol. 32, no. 1, pp. 96–101, 2018.
44. S. A. Osia, A. S. Shamsabadi, A. Taheri, H. R. Rabiee, and H. Haddadi, "Private and scalable personal data analytics using hybrid edge-to-cloud deep learning," *Computer*, vol. 51, no. 5, pp. 42–49, 2018.
45. N. Pratas, N. Marchetti, N. R. Prasad, A. Rodrigues, and R. Prasad, "Centralized cooperative spectrum sensing for ad-hoc disaster relief network clusters," in *2010 IEEE International Conference on Communications*, pp. 1–5, 2010.
46. W. Ejaz, N. Ul Hasan, and H. S. Kim, "Distributed cooperative spectrum sensing in cognitive radio for ad hoc networks," *Computer Communications*, vol. 36, no. 12, pp. 1341–1349, 2013.
47. A. Nadeem, M. Khan, and K. Han, "Non-cooperative spectrum sensing in context of primary user detection: A review," *IETE Technical Review*, vol. 34, no. 2, pp. 188–200, 2017.
48. S. Gmira, A. Kobbane, and E. Sabir, "A new optimal hybrid spectrum access in cognitive radio: Overlay-underlay mode," in *2015 International Conference on Wireless Networks and Mobile Communications (WIN-COM)*, pp. 1–7, 2015.
49. N. Muchandi and R. Khanai, "Cognitive radio spectrum sensing: A survey," in *2016 International Conference on Electrical, Electronics, and Optimization Techniques (ICEEOT)*, pp. 3233–3237, IEEE, 2016.
50. A. M. Koushik, F. Hu, and S. Kumar, "Intelligent spectrum management based on transfer actor-critic learning for rateless transmissions in cognitive radio networks," *IEEE Transactions on Mobile Computing*, vol. 17, no. 5, pp. 1204–1215, 2018.
51. W.-Y. Lee and I. F. Akyildiz, "Spectrum-aware mobility management in cognitive radio cellular networks," *IEEE Transactions on Mobile Computing*, vol. 11, no. 4, pp. 529–542, 2012.
52. H. Ye, L. Liang, G. Y. Li, J. Kim, L. Lu, and M. Wu, "Machine learning for vehicular networks: Recent advances and application examples," *IEEE Vehicular Technology Magazine*, vol. 13, no. 2, pp. 94–101, 2018.
53. A. Paul, A. Daniel, A. Ahmad, and S. Rho, "Cooperative cognitive intelligence for Internet of Vehicles," *IEEE Systems Journal*, vol. 11, no. 3, pp. 1249–1258, 2015.
54. P. H. Masur, J. H. Reed, and N. K. Tripathi, "Artificial intelligence in open radio access network," *IEEE Aerospace and Electronic Systems Magazine*, vol. 37, no. 9, pp. 6–15, 2022.

55. M. A. Hossain, R. Md Noor, K.-L. A. Yau, S. R. Azzuhri, M. R. Z'aba, I. Ahmedy, and M. R. Jabbarpour, "Machine learning-based cooperative spectrum sensing in dynamic segmentation enabled cognitive radio vehicular network," *Energies*, vol. 14, no. 4, p. 1169, 2021.
56. C. Chembe, D. Kunda, I. Ahmedy, R. M. Noor, A. Q. M. Sabri, and M. A. Ngadi, "Infrastructure based spectrum sensing scheme in VANET using reinforcement learning," *Vehicular Communications*, vol. 18, p. 100161, 2019.
57. X. Liu, C. Sun, M. Zhou, B. Lin, and Y. Lim, "Reinforcement learning based dynamic spectrum access in cognitive Internet of Vehicles," *China Communications*, vol. 18, no. 7, pp. 58–68, 2021.
58. R. Pal, N. Gupta, A. Prakash, R. Tripathi, and J. J. Rodrigues, "Deep reinforcement learning based optimal channel selection for cognitive radio vehicular ad-hoc network," *IET Communications*, vol. 14, no. 19, pp. 3464–3471, 2020.
59. K. Zhang, S. Leng, X. Peng, L. Pan, S. Maharjan, and Y. Zhang, "Artificial intelligence inspired transmission scheduling in cognitive vehicular communications and networks," *IEEE Internet of Things Journal*, vol. 6, no. 2, pp. 1987–1997, 2018.
60. Q. Huang, X. Xie, H. Tang, T. Hong, M. Kadoch, K. K. Nguyen, and M. Cheriet, "Machine-learning-based cognitive spectrum assignment for 5G URLLC applications," *IEEE Network*, vol. 33, no. 4, pp. 30–35, 2019.
61. X.-L. Huang, J. Wu, W. Li, Z. Zhang, F. Zhu, and M. Wu, "Historical spectrum sensing data mining for cognitive radio enabled vehicular ad-hoc networks," *IEEE Transactions on Dependable and Secure Computing*, vol. 13, no. 1, pp. 59–70, 2015.
62. R. Sarmah, A. Taggu, and N. Marchang, "Detecting byzantine attack in cognitive radio networks using machine learning," *Wireless Networks*, vol. 26, no. 8, pp. 5939–5950, 2020.
63. A. Taggu and N. Marchang, "Detecting byzantine attacks in cognitive radio networks: A two-layered approach using hidden Markov model and machine learning," *Pervasive and Mobile Computing*, vol. 77, p. 101461, 2021.
64. Z. Luo, S. Zhao, Z. Lu, J. Xu, and Y. Sagduyu, "When attackers meet AI: Learning-empowered attacks in cooperative spectrum sensing," *IEEE Transactions on Mobile Computing*, vol. 21, no. 5, pp. 1892–1908, 2020.
65. N. I. Mowla, N. H. Tran, I. Doh, and K. Chae, "Federated learning-based cognitive detection of jamming attack in flying ad-hoc network," *IEEE Access*, vol. 8, pp. 4338–4350, 2019.
66. L. Xiao, X. Lu, D. Xu, Y. Tang, L. Wang, and W. Zhuang, "UAV relay in VANETs against smart jamming with reinforcement learning," *IEEE Transactions on Vehicular Technology*, vol. 67, no. 5, pp. 4087–4097, 2018.
67. H. Gu, X. Guo, X. Wei, and R. Xu, "Mean-field controls with q-learning for cooperative marl: Convergence and complexity analysis," *SIAM Journal on Mathematics of Data Science*, vol. 3, no. 4, pp. 1168–1196, 2021.
68. Y. Yang, R. Luo, M. Li, M. Zhou, W. Zhang, and J. Wang, "Mean field multi-agent reinforcement learning," in *International Conference on Machine Learning*, pp. 5571–5580, PMLR, 2018.
69. A. Abouaomar, Z. Mlika, A. Filali, S. Cherkaoui, and A. Kobbane, "A deep reinforcement learning approach for service migration in MEC-enabled vehicular networks," in *2021 IEEE 46th Conference on Local Computer Networks (LCN)*, pp. 273–280, 2021.

Section 3

Protocol and Infrastructure

Chapter 3

Machine Learning Techniques in Conjunction with Data Science Applications for CR-IoV

Muhammad Nouman Noor, Saddaf Rubab, and Saad Rehman

3.1 INTRODUCTION

By the end of 2025, the enormous biological system of the Internet of Things (IoT) is anticipated to clear a plane mode for 0.1 trillion connections. In this way, the IoT can disrupt new companies [1]. The IoT is now moving towards the new domain known as the Internet of Vehicles (IoV) [2] because of the connection of smart transportation frameworks for improved application [3]. The IoV permits vehicles to speak with their inward and outside conditions. The correspondences of vehicles in exchanging data can be in an alternate structure. For instance, vehicles can exchange information with sensors, automobiles, and via the network [4]. The basic elements of the IoV are the connected vehicles. The IoV development is compelled by the unique versatile correspondence framework with capacities of social occasion, sharing, handling, registering, and getting the arrival of data [5].

IoVs along-with cognitive radio (CR) permit correspondence between vehicles in an arrangement of correspondence circumstances, extending the speed of data move and bandwidth. The use of CR can meet the future necessity for quicker data transport among vehicles and system (V2I). Vehicles along-with CR limits on IoV have a surprising appearance in contrast with ordinary IoV vehicles [6]. The pattern of CR is displayed in Fig-3.1. All vehicles are supposed to fight for channel-access on the 75 MHz range apportioned by the "US Federal Communication Commission (FCC)" in the 5.9 GHz range band for the WAVE framework and use it for the trading of wellbeing and infotainment data. Notwithstanding, to understand the maximum capacity of IoVs, smart vehicles should have the option to remotely trade correspondence with each other by means of vehicle-to-vehicle (V2V), vehicle-to-infrastructure (V2I), and vehicle-to-everything (V2X) interchanges [2]. It can be achieved by exploiting the large number of remote organizations and ranges, for example, cell and Wi-Fi organizations, TV groups, and satellite organizations, contingent upon their accessibility and the area of smart vehicles. Going against the norm, the expected expansion popular for different vehicular organization situated applications (security

DOI: 10.1201/9781003284871-6

and non-wellbeing related administrations) would unquestionably bring about a deficiency of phantom assets for IoV correspondence organizations.

As indicated by [2, 3], the arising Cognitive-Radio innovation has been imagined as an empowering idea with the possibility to beat the test of range shortage, which is the consequence of the current style of fixed spectrum allocation (FSA) strategy [4]. Dynamic spectrum access (DSA) or the range sharing component [5] has been decreed an indispensable potential related with CR innovation. Also, the current distribution of the range for specific radio advancements inside 300 MHz-3 GHz (i.e., the great recurrence groups) is drawing nearer to the immersion point. Hence, range distribution administrative bodies like the "UK Office of Communications (Ofcom)" or the US FCC are thinking about more adaptable range the board systems, for example, the auxiliary range access instrument [2]. Accordingly, the plan and improvement of exceptional novel radio advances, for example, DSA or the range sharing component [5, 6] are imperative, to have the option to direct activities in unlicensed groups.

As our surroundings become more associated as a rule, Intelligent Transportation Systems will assume a focal part in our urban communities and across borders, framing part of another vision of "versatility as a help". Connected vehicle innovation, with a conspicuous job in Intelligent Transportation Systems, will be equipped for creating colossal measures of unavoidable and continuous information, at extremely high frequencies. These streaming information are the normal kind of information created by associated vehicles, and their examination is of vital significance for applications further developing street security, powerful help conveyance, eco-driving, traffic guideline and contamination decrease. In this chapter, we are targeting to describe this sort of information according to a scientific viewpoint, as well as to represent the difficulties information science faces in separating information from them progressively. Information produced by sensors and actuators in associated vehicles incorporate loud, atypical, excess, quickly changing, connected and heterogeneous information. In such a unique circumstance, various methods have been proposed to adjust the information examination in bunch figuring out how to these new powerful and developing streaming information, which are created in tremendous volumes and sent at high speed. The Internet of Vehicles can possibly give an unavoidable organization of associated vehicles, smart sensors and street foundations, and huge information can possibly process and store that measure of information and data. Demonstrating, anticipating, and removing significant data in sensible and proficient ways from large information address a test for data science in associated vehicles.

Artificial Intelligence, the sub branch of which is Machine Learning (ML) as well as Data Science (DS) is a computerized reasoning techniques used to show a framework the obscure and make proficient and compelling choices. The utilization of ML and data sciences in nearly all angles, like mechanical technology, business, arts, automated frameworks, biotechnology, and

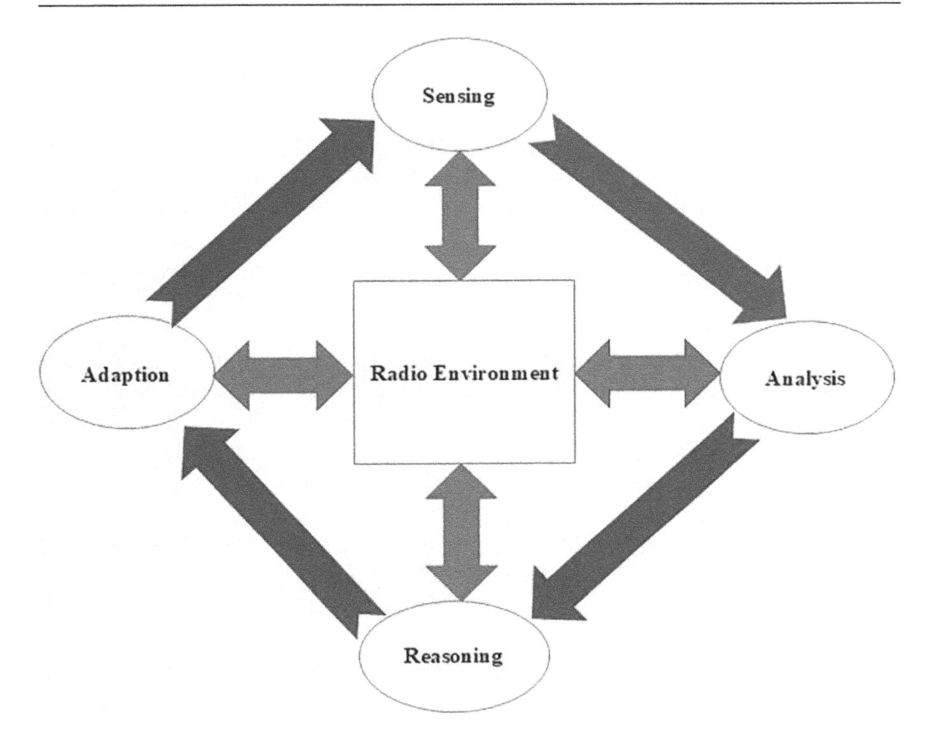

Figure 3.1 Cycle of Cognitive Radio

intelligent automated transportation frameworks, has become well known due to the accessibility of minimal expense and exceptionally competent (i.e., high-computational-power) [7] machines and the existence of enormous measures of information. DS and ML provide shrewd and rapid decision making for further developing system efficiency that includes dependability, energy productivity, and quality-of-service (QoS) [7].

The outstanding development of remote gadgets has prompted the requirement for an immense range to help high-volume information transmission. Notwithstanding, range shortage (insufficient portion contrasted and the interest) has turned into a block to the sending, backing, and scaling of cutting edge solicitations for workers, including the IoT, smart urban communities, computer generated reality, increased reality, and top quality 3D video web-based features. The two fundamental factors that cause range shortage are per the following: (a) recurrence groups are distributed to authorized clients in view of the conventional fixed range task strategy and (b) an enormous volume of constant information is created and moved over a remote medium in a unique climate. The creators in [8] showed that most groups are as yet empty and appropriate for auxiliary use.

The main contributions of this chapter are:

- Thorough concepts of machine learning, IoV and data science along with their associations are presented.
- The applications and usages of data science and machine learning in CR-IoV are discussed.

3.2 THE COMMUNICATION OF CR-IoV

IoV is a mix of three organizations: an inter-vehicle linkage, an intra-vehicle linkage, and vehicular transportable internet. In light of this idea of three organizations incorporated into one, we characterize an IoV as a huge scope conveyed framework for remote correspondence and data trade between V2X (in which X is everything else like street, web and vehicle), per harmonized communication contracts and data communication strategies (models incorporate the IEEE 802.11p WAVE standard, and possibly cell innovations). It is an incorporated organization for supporting smart traffic the board, smart unique data administration, and smart vehicle control, addressing a normal utilization of IOT innovation in intelligent transportation framework (ITS). The union of innovation includes data correspondences, natural security, energy preservation, and wellbeing. To prevail in this developing business sector, obtaining of center advancements and guidelines will be significant to secure an upper hand. In any case, the joining of the IoV with other frameworks ought to be pretty much as significant as the structure of the IoV progresses itself. As a result of this, the IoV will turn into a basic piece of the biggest IOT framework by its finish. Here, it should be stressed as vital, that coordinated effort and inter communication between the transference area and different areas (for example, energy, medical care, climate, assembling, and horticulture, etc. . . .) will be the subsequent stage in IoV improvement.

The incredible upheaval that raised from the Internet has given an open door to interfacing individuals at an excellent greatness and speed. The achievement recorded from the Internet unrest achieved huge open door that is as of now changing the strategies by which different items impart as of now. This fast improvement contemplates the inter-connection between elements to understand a savvy conurbation, where a gadget interfaces with other associated gadgets. This correspondence is accomplished through consistent pervasive detecting, arising innovations, and accessibility of a versatile stage for enormous information investigation. As of now, objects, for example, cell phones, vehicles, PCs and tablets and other palmtop gadgets, change our environmental elements, making them extremely intuitive and enlightening [9, 10]. Through present day correspondence, shrewd gadgets make an organization of interconnected objects with ongoing communications. The development in the quantity of gadgets and the idea of the worldwide organization design, which incorporates all current heterogeneous organizations, has formed our understanding. This widespread organization of possessions has been recognized as an imminent Internet as of now molded as the IoT [**11**]. The IoT fills in as an empowering climate where sensors and actuator-objects connect flawlessly and give logically more appropriate stages to information trade. The new headway and transformation of different remote correspondence innovations have situated IoT on the way to be an auspicious innovation, which doles from the prospective possibilities gave through Internet innovation. The IoT innovation has achieved the improvement of astute frameworks, which incorporate yet are not restricted to shrewd retail, smart water, savvy energy, shrewd matrices, smart medical services, savvy homes, and smart transportation [**9, 10**]. The IoT has made interfaces for savvy gadgets to be associated with a worldwide organization with the capacity to deliver administrations from other associated gadgets [**12**]. The IoT empowers consistent mix of heterogeneous organization of gadgets using keen connection points. In this manner, one of the critical goals of the IoT is interoperability between dissimilar gadgets [**9, 11–13**]. The rise of IoT innovation has altered many new innovative work regions.

The worldwide vehicular traffic was anticipated to heighten to three hundred thousand Exabytes around the end of 2025. This huge expansion in vehicular information is the outcome of the headway in vehicular-telematics applications, remembering for in-vehicle infotainment (IVI) and ITS [**14**]. Traditional VANETs utilize automobile as a hub for sending or handing-off traffic data among vehicles and frameworks utilizing V2V and V2I correspondences. Numerous vehicular applications, including ITS administrations and wellbeing application, have utilized the possibilities presented by the rising availability of current vehicles. For example, V2V correspondence empowers dividing of data between vehicles for security correspondence spread. Alternately, V2I correspondence empowers the assortment of data from various framework offices [**14**]. The pictorial depiction of communication in a CR-IoV is shown in Fig-3.2.

This scientific categorization presents the various kinds of associations that exist among vehicles and different gadgets. Moreover, it recognizes the data stream in each IoV correspondence classification as well as the arising advancements used by every correspondence type. The kind of the correspondence included "vehicle-to-vehicle interchanges", "vehicle-to-vehicle correspondences", vehicle-to-infrastructure (V2I)", "vehicle-to-sensor (V2S) interchanges", "vehicle-to-personal gadget (V2P): interchanges", "vehicle-to-pedestrian (V2D) correspondences", and "vehicle-to-home (V2H) correspondences".

3.3 DEMAND OF MACHINE LEARNING AND DATA SCIENCE IN CR-IoV

Various new vehicles are supposed to show up on streets before very long, and they would cause serious gridlock that can deaden metropolitan regions

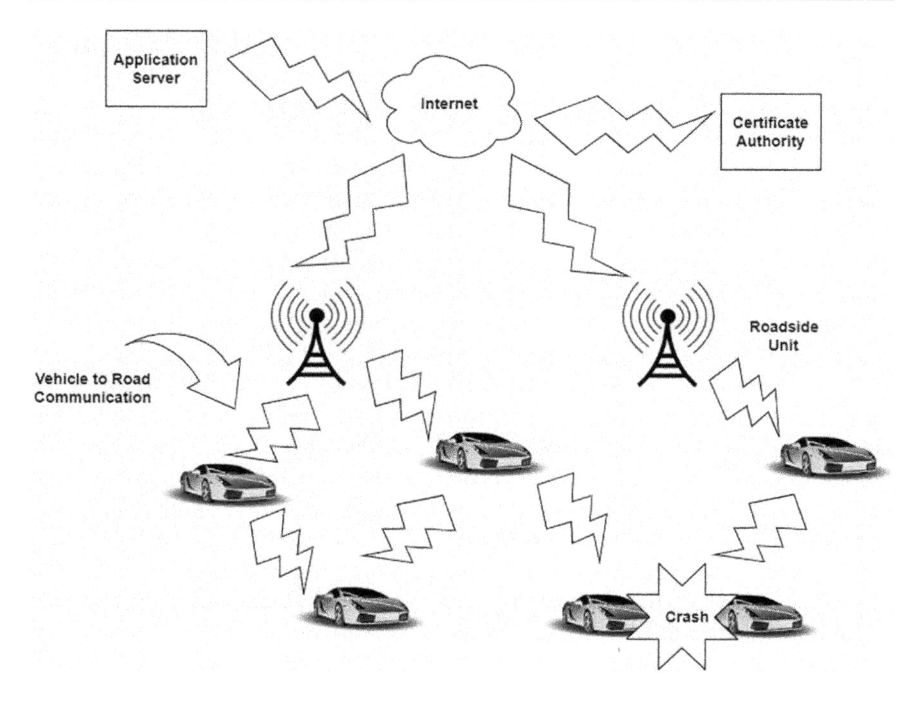

Figure 3.2 Communication in a CR-IoV

and unfavorably influence the economies of nations. Aside from adding to monetary misfortunes, unfortunate administration of transportation frameworks can make pressure individuals, diminish working effectiveness, and increment the quantity of mishaps and losses. To address these concerns, the shrewd transportation framework or IoV should be enhanced to get a mechanized smart traffic framework that gives helpful data on street and traffic circumstances and autonomous vehicles (AV) [15]. For an effective execution of IoV, a monstrous measure of live information should be traded. As indicated by Intel, a peculiarity called "flooding of data" is supposed to happen, by which each savvy AV would create and consume roughly four terabytes of information on the normal each diurnal of driving. This sum is ordinarily bigger than the ongoing measure of information that a typical individual right now creates.

An effectively working vehicle can create a measure of information that 3000 individuals right now produce overall. This information, which can be assembled by sensors, cameras, and publicly supporting, incorporate street and traffic conditions, individual information, and application information (e.g., advertising, cultural, and diversion information). Hence, information is the next "oil" in the transportation framework. Notwithstanding, the transmission capacity expected to oblige such enormous constant information trade is scant, bringing about network clog, particularly in metropolitan regions.

In the CR framework, an un-licensed client (SU) distinguishes any empty or vacant authorized recurrence claimed by authorized clients (PUs). Upon distinguishing an empty recurrence band, the SU is permitted to use it furnishing that it disrupts no PU. Consequently, the SU should deliver the recurrence band when the PU's exercises return. The SU should guarantee that its communication power doesn't slow down the PU's exercises in the vicinity [**16**]. The data attainment from IoV is shown in Fig-3.3.

Fast versatility and a unique climate have achieved extra intricacies and difficulties to CR-IoV. ML and data science strategies can ease these intricacies and give colossal upgrades regarding network execution upgrade (e.g., diminished delay, expanded dependability, secure execution, furthermore, energy productivity) to CR-IoV. Albeit the energy limit of vehicles is by and large adequate, the aggregate energy prerequisite of vehicles can be extremely high; in this manner, energy proficiency should be accomplished in light of the enormous fossil fuel byproduct that can represent a danger to the green climate [**17**]. Another significant issue is to move along the QoS and quality-of-experience (QoE) of the organization since the regular range detecting, transmission transformation, and handover in the CR framework increment the postponement, above, and energy utilization [**18**] ML likewise gives an ideal course to CR-IoV clients to keep away from gridlocks and street mishaps. ML can likewise assume an essential part in the best

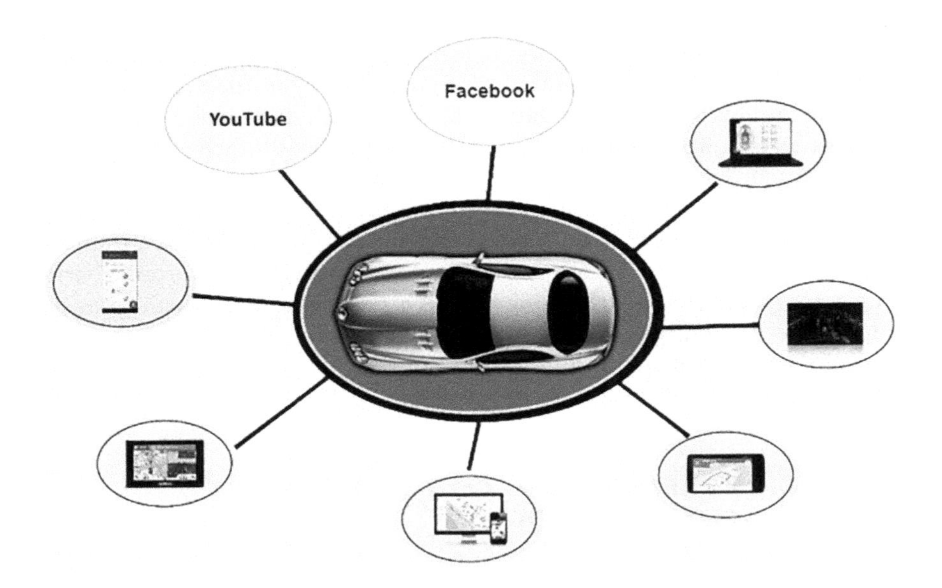

Figure 3.3 Data Capturing from IoV

infotainment practice in CR-IoV. It tends to be utilized for fitting booking, choosing the best channel, and focusing on messages.

CR and ML can assume a significant part in the future driverless vehicle framework. The job of CR in the future mobilization framework has been introduced in past conversations. This chapter demonstrates the way that ML can be smeared to diminish street mishaps and gridlock. CR can be used to oblige the range expected to help monstrous information correspondence among mechanized AV and organizations. ML can be an indispensable piece of this self-sufficient or mechanized vehicle framework. Like a robot, an AV can gain proficiency with the encompassing climate and speak with expanded well-being, dependability, QoS, and energy productivity by smearing such learning.

3.4 UTILIZATIONS OF DATA SCIENCE TO CR-IoV

The effects and expected functional advantages of the utilization of DS to IoVs might be characterized on versatility of individuals, safety, ecological advantages and driver support. The accompanying viewpoints might become achievable by the insight got from DS. The vast majority of them depend on unmistakable, demonstrative or predictive analytics i.e. Agility, Security and Support.

In terms of Agility, the main utilization is to distinguish and figure out examples and patterns in versatility information and other utilizations are changing traffic lights for progressively overseeing travel tasks, or for crisis directing, traffic administration during arranged or spontaneous occasions, Decrease primary city streets' clog (and accordingly air pollution) by anticipating traffic stream and Carpooling proposals. For security, its applications are variable speed-limit frameworks for guaranteeing traffic cautious, driver conduct and execution investigation, to distinguish ill driving occasions by various causes, identify critical components of the IoV, help the driver in ideal IoV activity, and in this way, expanding asset economy and vehicle lifetime, surmise constant ecological circumstances per IoVs gathered information, path changing help, foresee the impact of natural circumstances detected by the IoV on vehicle inhabitants and distinguishing proof of drivable street surface and street limits. In terms of support, the main usage is supportive versatile journey control, in transit direction to parking spots, capacity of IoVs clients to find area based data of interest, driver conduct investigation and vehicle prescient upkeep.

3.5 MACHINE LEARNING IN CONJUNCTION WITH DATA SCIENCE FOR CR-IoV

Completely independent, semi-autonomous as well as traditional vehicles are currently operating in the IoV. IoV can uphold large data acquirement, retrieval, broadcast, and processing. The data produced by the IoV can improve the effectiveness of the network and its performance. [19]. This data is mostly unstructured in nature as it contains images, videos, audios, therefore it is difficult to analyze the using traditional and statistical techniques hence requires Machine Learning (ML) for its processing. It is shown in different researchers that ML and its sub branch Deep Learning (DL) are effective for processing this type of data [20]. ML is a famous device for enormous amount of unstructured data [21, 22] due to its extraordinary outcome in various fields [23].

The ML architecture is complicated with the ability to chip away at enormous information produced from the IoV. These ANN works better compared to the basic and traditional statistical techniques [24]. ML has shown propitious accomplishment in unstructured information examination. The promising exhibition of ML in handling unstructured information, for instance, in visual article classification, NLP, and data recovery, has been accounted for in the writing [25].

On the account of IoV, ML is very useful and must entail on the grounds as it learns on the basis of experience. As the matter of fact is that practically all suitable potential outcomes are completely robotized, ML is expected to catch new situations and perform investigation on the gathered data from different IoV devices like cameras and sensors. Therefore, this approach is very helpful in the IoV surroundings to stay protected and reduce the death rate because it enables the vehicle to take critical decision at right time. Adequate preparation objects are given when large information are taken advantage of. Subsequently, the presentation of ML is gotten to the next level. Preparing of enormous scope ML models for huge information include requires elite execution frameworks and design, like GPUs and heavy CPUs [26].

For CR-IoVs ML and DS can be used for enhancing road traffic safety through the collaboration of inter-vehicle, intra-vehicle, and beyond-vehicle networks. The intelligent autonomous moving object (represented by the autonomous vehicle) is the most significant component of CR-IoV and has the appropriate perception and comprehension capabilities for its surroundings as well as the capacity for automatic decision-making. Within the boundaries of the intra-vehicle network, the unmanned vehicle should be able to identify the driver's driving style. If autonomous vehicle production proceeds as planned, people and unmanned vehicles will cohabit for an extended period of time. As a result, it will be crucial to study the mechanism that allows an unmanned vehicle and a driver to cooperate while driving. The aspect to be taken into account in the cognitive IoV is each driver's current driving situation. Also predicting pedestrian behavior is the most crucial category in the obstacle behavior prediction in the beyond-vehicle network environment. The goal of the pedestrian detection and behavior prediction in the convoluted road at identifying the nearby pedestrian target circumstances, dividing into categories based on traits such target's histogram of color, texture, and gradients, as well as its grey level and edge identification of the pedestrian target, analysis of the pedestrian's height, age, and other data, and prediction of the target's risky actions are all part of the follow-up cognitive process. There are numerous studies concentrating on the effectiveness of classification algorithms, such as [27] combining the "discrete hidden markov model (DHMM)" and decision tree for differentiating behavior. Another main concern for CR-IoV is network security protection. In order to ensure the validity of model training and introduce the private feature encryption mode of the owner (such as basic biological feature and habitual driving behavior of the owner) on this basis, it makes use of the semi-supervised learning algorithm and takes into account the characteristics of privatization and diversity of the intra-vehicle network. This allows it to realize the precise identification of attack. It also makes it possible to forecast a dangerous route when used in conjunction with a joint examination of physical and network space. The resource cognitive engine executes the implementation monitoring for the network flow in the network space. The data cognitive engine performs cognitive analysis on the network flow that is fed back from the resource cognitive engine and provides real-time feedback for automatic driving in conjunction with joint analysis of perception data from the vehicle to surrounding road conditions, driving data from nearby vehicles, and data gathered from the intelligent traffic system (traffic network density and vehicle moving state in the physical space). The sensitivity of network data security will be quickly increased if an unusual driving behavior of a vehicle is identified, and the network space problem will be quickly identified and fixed. In the portion of Open Issue after this one, we will discuss more network security concerns and offer many potential remedies [28–30].

A new report has revealed that transformative ANN has many usages in the IoV. Chen et al. [31] exhibited that the developmental ANN can foresee backside crash inside the IoV surroundings. Thus, it can help in the advancement of a viable backside impact identification framework for automobiles in IoV conditions. The authors' in [32] projected the utilization of data science for the expectation of diminutive traffic stream. The chapter is roused by the collection of enormous information in the IoV, and the shallow counterfeit brain network calculation can't deal with such a lot of information. The data science is smeared for the diminutive traffic stream, and it accomplishes better compared to the pattern calculations. In the research paper [33], authors' showed a ML model for ideal responsibility designation to further develop vehicle vitality utilization in the IoV. ML gives upgraded energy proficiency and works on the inactivity of the organization. Ning et al. [34] explained CNN to work on the promptness of information broadcast and improve the data-sharing among automobiles in the IoV surroundings. The ConvNet is smeared for information transmission by taking advantage of the tri-connection between automobiles. The outcome demonstrates the effectiveness of the anticipated CNN in light of dormancy, message conveyance, and level of associated gadgets.

In [34], authors' proposed "hybridized motif based method (MBM)" and "ConvNet (MBM-ConvNet)" for "D2D" correspondence in the IoV. MBM bunches the smart cell phones in transports and with travelers in a three-sided way though the CNN envisages the D2D association. The "MBM-ConvNet" model shown improved results compared to the pair disclosure conspires, social mindful methodology, and "MBM". The concern now is that edge organization may not be great for crises. The research proposed in [35], hybridized "energy assessment conspire (EES)", "Wiener process model (WPM)", and "ConvNet (EES-WPM-ConvNet)" to guarantee upgraded throughput and decreased dormancy for information broadcast in the IoV. The EES inspects vehicle's energy stage for network by contrasting it and limit esteem, WPM gauges vehicles availability while CNN envisages the ultimate vehicle matches for information transmission. The proposed "EES-WPM-ConvNet" shown better results compared to the "EES-WPM". The test is that network is permitted exclusively for automobiles with adequate measure of energy. In the research paper [34], authors' anticipated hybridized "genetic algorithms (GA)" and "CNN" with "simulated annealing (SA)" calculation "(GA-ConvNet with SA)" to guarantee diminished energy level utilization for the two automobiles and street-sign element in the IoV. GA chooses the server with least power utilization for handling the lining demands, SA looks for the worldwide ideal answer for the introduction period of the CNN, and the CNN predicts the ideal responsibility portion for the computational offices. The outcome showed that the GA consumes less power contrasted with the CNN and SA while the CNN has the least organization delay contrasted and the GA and SA. Notwithstanding, the paper expected that all automobiles continue on a straight lane. The researchers' in [34] hybridized "Edmonds-Karp Algorithm (EKA)" and "DRL" with profound "Q-organization (DQN) (EKA-DDQN)" to limit how much energy spent in computational relieve in the environment of IoV. The "EKA" guarantees stream rerouting among "RSUs" while the "DDQN" limits the general energy utilization. The outcome acquired exhibited that the "EKA" achieved superior results compared to the ravenous technique and comprehensive strategy while the "DDQN" outflanked "Q-learning" and cloudlet processing models. That's what the issue is assuming the pace of information offloads is over the computational ability of the vehicles, say 80MB, the pace of energy utilization increments quickly.

It is evident from researchers that by analyzing and performing predictive analytics (DS and ML) on the data produced by IoV, various tasks can be performed like monitoring the health of vehicle, valuation on changing of lanes, recognition of traffic signals as well as road signs, smartly parking vehicle, advanced toll collection and predicting the flow of traffic [35–38]. Various applications of Data Science (DS) in conjunction with ML to ensure road safety are shown in Fig-3.4 and Fig-3.5 shows the applications of ML in reduction of traffic congestion [39].

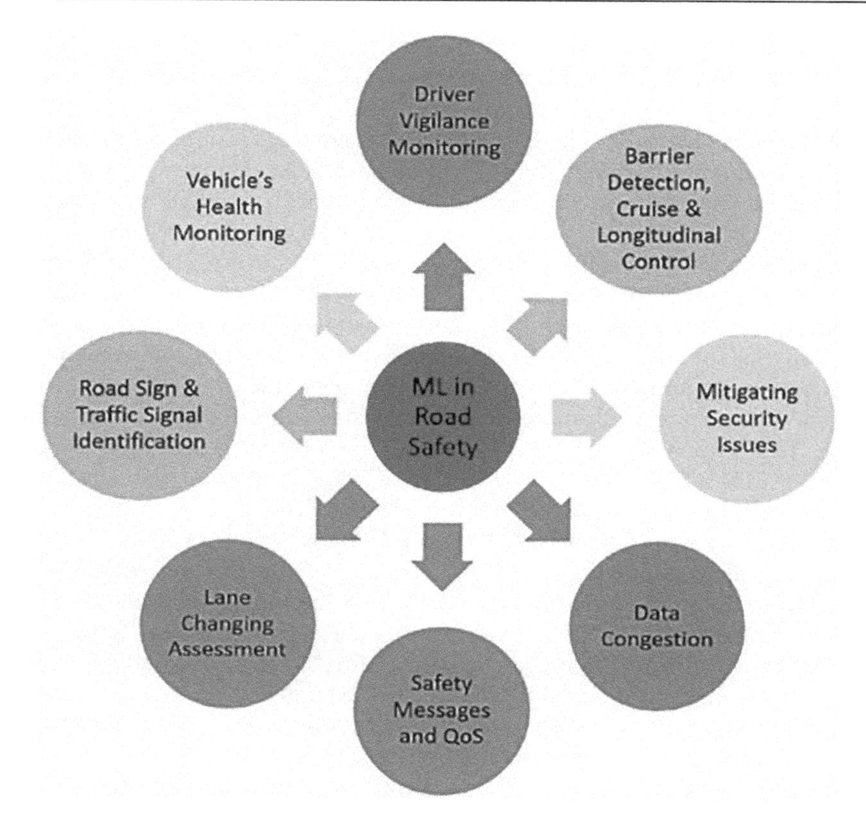

Figure 3.4 Various applications of ML and DS in Road Safety

Other AI procedures like the "fuzzy-systems", "random-forest", and "k-nearest neighbor" ordinarily witness decaying execution as how much information increments, which makes them unsuitable for data analysis. As talked about in Ali et al. [40], support vector machine has the test of managing quick verification system for huge scope IoV engineering. Fluffy framework has the limit of managing IoV interactive media correspondences. Shallow calculations like the "Random Forest", "multi-facet perceptron", and "AdaBoost" are confronting the test of getting choice for security in the "V2X traffic".

What's more, other AI procedures require separate strategies for highlight extraction prior to taking care of the information to the calculation for handling, which increments computational expense and requires human intercession, while ML has implanted programmed include extraction component that causes the DL calculation to kill the prerequisite for additional element extraction methods, in this way lessening the work of information designing. Subsequently, it gives ML advantage in predictive data analytics over other AI methods. It is notable in the writing that ML engineering, explicitly CNN, has demonstrated to be remarkable in picture handling contrasted with other AI procedures. DL enjoys the benefit of managing normal unlabeled information better than other AI strategies.

Moreover, the utilization of DL in DS has the accompanying qualities: capacity to create characteristic elements, compelling handling of unlabeled information, high precision in giving outcomes, and productivity with multimodal information [**41**]. The authors' examine it with regards to the IoV. Achieving high precision and accuracy in the environment of IoV is an urgent concern on the grounds that the automobiles in the IoV climate rely upon the choice of the ML framework. Exact examination can forestall bedlam on the public streets that can prompt mishaps, injury, and conceivably passing. For instance, erroneous catching of new situation by the ML framework could cause a deadly mishap in the environment of IoV.

The 3D guide information is chronicled by the robotized driving guides. Inside the space of a couple of centimeters (CMs) away, the 3D guide information is exact for the automobile position. The automobile identifies and follows different automobiles with an elevated degree of exactness, perceives paths, and measures distance and speed. This situation regularly happens when the item and natural innovation of the vehicle is empowered [**42**]. The ML framework assumes a huge part in this situation. The sensors implanted in the automobiles inside the environment of IoV create information with

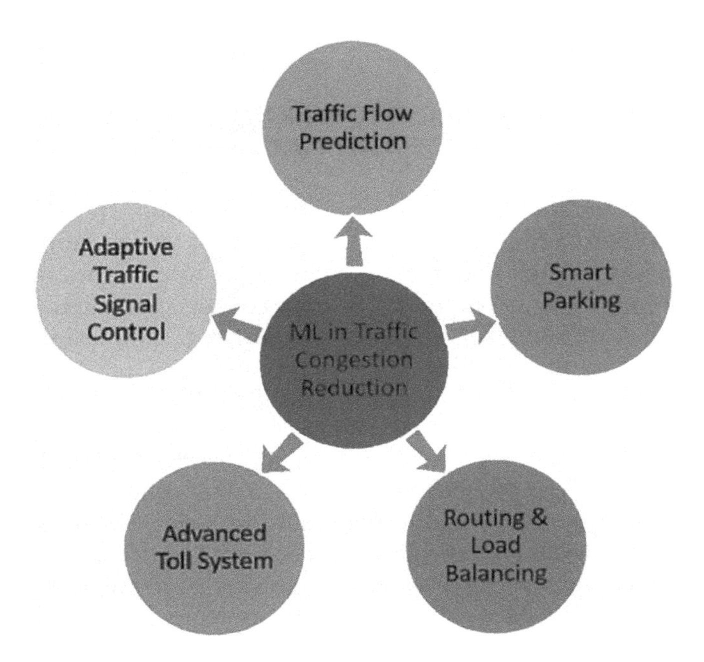

Figure 3.5 ML and DS Application in Reduction of Traffic Congestion

characteristic element in light of the fact that the information are acquired from the sensors. Data science requires inherent highlights, and ML can create the natural elements expected by data analytics. The element is a trait of sensor information. Significant level elements can be advanced naturally by ML without manual intercession. A huge piece of the information produced from the sensors implanted in automobiles in the environment of IoV alludes to normal information. Unique in relation to traditional AI procedures that require huge designing works, DL can successfully manage normal unlabeled information with negligible human mediation. Accordingly, human exertion in marking information is limited. The sensors create different information (pictures, sound, and discourse), and DL can work with multimodal input information.

Deep reinforcement learning [43] can likewise assume an imperative part inside the IoV conditions in view of the intricacy of certifiable driving. In independent vehicles, ML [44], elite execution processing framework, and high level calculations are expected for the automobiles to adjust to evolving circumstances. This methodology can be accomplished through 3D superior quality guides. The cameras and sensors in the independent vehicles produce huge scope information for aggregation. The information is expected to be examined to keep the automobile continuing on the path. Without ML that utilizes the data from superior quality guides that contain geocoded information, completely independent driving turns into a hallucination. Without top quality guides containing geocoded information and ML that utilizes this data, completely independent driving deteriorates in Europe [42].

3.6 PATTERNS IDENTIFICATION FROM CR-IoV DATA USING DATA SCIENCE AND MACHINE LEARNING

The pattern identification process for old style information revelation is usually parceled into a few phases. In the information stream setting, the pattern identification cycle should be thought again to deal with information from sensor networks in real-time. The steps contained in generic system are consisted of Acquiring Data, Preprocessing of Data, Data fusion, Analytics and Prediction as shown in Fig-3.6.

Data which are coming through CR-IoV are heterogeneous in nature therefore it is a challenging task. When a data is acquired from different sensors, its complexity reduction is an obligatory stage. This reduction attempts to keep up with the first design and significance of the data sources, and yet getting a significantly more reasonable size. Quick training and further developed speculation capacities of learning calculations, as well as better comprehension and interpretability of results, are among the many advantages of data-reduction. A few strategies have been produced for putting away outlines or summary data about recently seen information (for example outlining calculations, component counting information

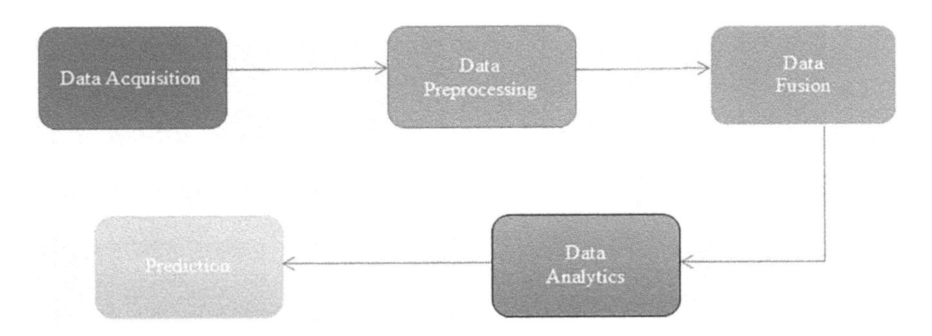

Figure 3.6 Generic Methodology for Patterns Identification

structures). Frequently, the interest is to process some measurable property of the streaming information. The recursive adaptation of the sample mean or incremental version of the standard deviation can be utilized, to refer to a couple, to get basic measurements of information streams. Likewise, Hoeffding bounds give certainty limits on the mean of a conveyance (offered an adequate number of free perspectives, the genuine mean of an irregular variable won't vary from the assessed mean by in excess of a specific sum). The primary interest of these insights is that they permit keeping up with their accurate qualities over an in the long run endless grouping of information without putting away them in memory. Numerous various types of abstract can be developed relying on the application within reach. The idea of the abstract exceptionally impacts the kind of bits of knowledge that can be mined from it.

After data is acquired, preprocessing is performed and it is the stage that incorporates tasks to readies the information for further investigation and work on the nature of the data. The heterogeneity in pre-handling tasks emerge as the need should arise handle missing qualities, eliminate outliers, and distinguish anomalies and remove noise from the data. Then data fusion phase comes in which the data which is receiving from different devices are fused in a single form in order to produce a better result that a single data source can produce. Different data fusion forms are complementary fusion, redundant fusion and cooperative fusion. Also other techniques like data aggregation, serial and parallel fusion, data combination and sensor fusion can also be performed for the said purpose.

Then data analytics is performed to find the patterns in the data using different DS techniques like Decision Trees, Support Vector Machine, Resampling, linear regression and bootstrapping. These techniques help to find and gain the insights of data which may be helpful for different applications like traffic congestion, driver's monitoring, vehicle health monitoring and many others. At last model is trained for predictions in future. The complete learning process is shown in Fig-3.7.

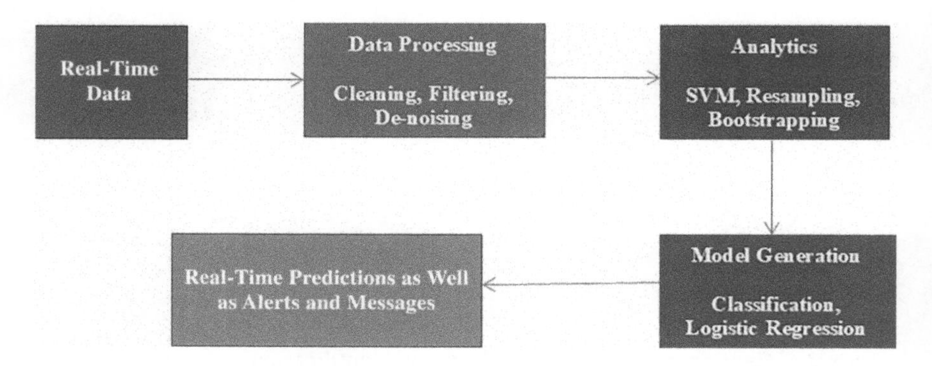

Figure 3.7 Complete Learning Process

3.7 CONCLUSION

In this chapter, we presented a study on applications of machine learning in the context of CR-IoV in conjunction with data science. The link that exists amid Machine Learning, IoV, and data science has been revealed to give specialists a reasonable point of view on the experimental utilization of ML with data science in the CR-IoV. The outcomes show that exact chips away at ML in the environment of IoV are exceptionally restricted and public vault information for IoV are mostly inaccessible to researchers. We assent that this study can assist scientists with effectively distinguishing regions that require arrangements and fledgling researchers can encompass it as a point of reference.

REFERENCES

1. Huawei Technologies Co, *Huawei*, Huawei, Shenzhen, China, 2015. www.huawei.com/en/.
2. IEEE, *IEEE Transmitter*, IEEE, Piscataway, NJ, USA, 2018.
3. E. S. Ali, M. K. Hasan, R. Hassan et al., "Machine learning technologies for secure vehicular communication in Internet of Vehicles: Recent advances and applications," *Security and Communication Networks*, vol. 2021, p. 23, Article ID 8868355, 2021.
4. N. Lu, N. Cheng, N. Zhang, X. Shen, and J. W. Mark, "Connected vehicles: Solutions and challenges," *IEEE Internet of Things Journal*, vol. 1, no. 4, pp. 289–299, 2014.
5. N. Liu, "Internet of vehicles: Your next connection," *Huawei WinWin*, vol. 11, pp. 23–28, 2011.
6. M. Arif, V. D. Kumar, L. Jayakumar, I. Ungurean, D. Izdrui, and O. Geman, "DAHP–TOPSIS-based channel decision model for co-operative CR-enabled internet on vehicle (CR-IoV)," *Sustainability*, vol. 13, no. 24, p. 13966, 2021. https://doi.org/10.3390/su132413966.

7. K.-L. A. Yau, P. Komisarczuk, and P. D. Teal, "Applications of reinforcement learning to cognitive radio networks," 2010 IEEE International Conference on Communications Workshops, pp. 1–6, 2010.
8. D. Das and S. Das, "A survey on spectrum occupancy measurement for cognitive radio," *Wireless Personal Communications*, vol. 85, no. 4, pp. 2581–2598, 2015.
9. J. Gubbi, R. Buyya, S. Marusic, and M. Palaniswami, "Internet of things (IoT): A vision, architectural elements, and future directions," *Future Generation Computer Systems*, vol. 29, no. 7, pp. 1645–1660, 2013.
10. I. A. T. Hashem, V. Chang, N. B. Anuar et al., "The role of big data in smart city," *International Journal of Information Management*, vol. 36, no. 5, pp. 748–758, 2016.
11. O. Kaiwartya, A. H. Abdullah, Y. Cao et al., "Internet of vehicles: Motivation, layered architecture, network model, challenges, and future aspects," *IEEE Access*, vol. 4, pp. 5356– 5373, 2016.
12. Elsaddik, "Multimedia communications research laboratory," 2016. www.mcrlab.net/research/social-networks-for-the-internet-of-vehicles/.
13. F. Yang, S. Wang, J. Li, Z. Liu, and Q. Sun, "An overview of Internet of Vehicles," *China Communications*, vol. 11, no. 10, pp. 1–15, 2014.
14. W. Xu, H. Zhou, N. Cheng et al., "Internet of vehicles in big data era," *IEEE/CAA Journal of Automatica Sinica*, vol. 5, no. 1, pp. 19–35, 2017.
15. X. He, W. Shi, and T. Luo, "Survey of cognitive radio VANET," *KSII Transactions on Internet and Information Systems*, vol. 8, no. 11, pp. 3837–3859, 2014.
16. I. F. Akyildiz, W. Lee, M. C. Vuran, and S. Mohanty, "A survey on spectrum management in cognitive radio networks," *IEEE Communications Magazine*, no. April, pp. 40–48, 2008.
17. R. Atallah, C. Assi, and M. Khabbaz, "Deep reinforcement learning-based scheduling for roadside communication networks," 2017 15th International Symposium on Modeling and Optimization in Mobile, Ad Hoc, and Wireless Networks (WiOpt), pp. 1–8, 2017.
18. A. K. Sadek, K. J. R. Liu, and A. Ephremides, "Cognitive multiple access via cooperation: Protocol design and performance analysis," *IEEE Transactions on Information Theory*, vol. 53, no. 10, pp. 3677–3696, 2007.
19. W. Xu, H. Zhou, N. Cheng et al., "Internet of vehicles in big data era," *IEEE/CAA Journal of Automatica Sinica*, vol. 5, no. 1, pp. 19–35, 2017.
20. A. Samuel, M. I. Sarfraz, H. Haseeb, S. Basalamah, and A. Ghafoor, "A framework for composition and enforcement of privacy-aware and context-driven authorization mechanism for multimedia big data," *IEEE Transactions on Multimedia*, vol. 17, no. 9, pp. 1484–1494, 2015.
21. M. A. Alsheikh, D. Niyato, S. Lin, H.-P. Tan, and Z. Han, "Mobile big data analytics using deep learning and Apache spark," *IEEE Network*, vol. 30, no. 3, pp. 22–29, 2016.
22. K. Yu, "Large-scale deep learning at Baidu," Proceedings of the 22nd ACM International Conference on Information & Knowledge Management, San Francisco, CA, USA, 2013.
23. J. P. Papa, G. H. Rosa, D. R. Pereira, and X.-S. Yang, "Quaternion-based deep belief networks fine-tuning," *Applied Soft Computing*, vol. 60, pp. 328–335, 2017.

24. N. Straton, R. R. Mukkamala, and R. Vatrapu, "Big social data analytics for public health: predicting Facebook post performance using artificial neural networks and deep learning," Proceedings of the 2017 IEEE International Congress on Big Data (BigData Congress), Honolulu, HI, USA, 2017.
25. J. Dean, G. Corrado, R. Monga et al., "Large scale distributed deep networks," Proceedings of the 2012 Advances in Neural Information Processing Systems, Lake Tahoe, NV, USA, 2012.
26. C. Zhang, K. C. Tan, H. Li, and G. S. Hong, "A cost-sensitive deep belief network for imbalanced classification," *IEEE Transactions on Neural Networks and Learning Systems*, vol. 30, no. 1, pp. 109–122, 2018.
27. S. Reddy, M. Mun, J. Burke, D. Estrin, M. Hansen, and M. Srivastava, "Using mobile phones to determine transportation modes," *ACM Transactions on Sensor Networks*, vol. 6, no. 2, pp. 1–27, 2010.
28. M. Chen, Y. Tian, G. Fortino, J. Zhang, and I. Humar, "Cognitive Internet of Vehicles," *Computer Communications*, vol. 120, pp. 58–70, 2018.
29. H. Chang, Y. Liu, and Z. Sheng, "Blockchain-enabled online traffic congestion duration prediction in cognitive internet of vehicles," *IEEE Internet of Things Journal*, vol. 9, pp. 25612–25625, 2022.
30. X. Liu, C. Sun, K.-L. A. Yau, and C. Wu, "Joint collaborative big spectrum data sensing and reinforcement learning based dynamic spectrum access for cognitive internet of vehicles," *IEEE Transactions on Intelligent Transportation Systems*, vol. 25, pp. 805–815, 2022.
31. C. Chen, H. Xiang, T. Qiu, C. Wang, Y. Zhou, and V. Chang, "A rear-end collision prediction scheme based on deep learning in the internet of vehicles," *Journal of Parallel and Distributed Computing*, vol. 117, pp. 192–204, 2018.
32. F. Kong, J. Li, B. Jiang, and H. Song, "Short-term traffic flow prediction in smart multimedia system for Internet of Vehicles based on deep belief network," *Future Generation Computer Systems*, vol. 93, pp. 460–472, 2019.
33. X. Wang, X. Wei, and L. Wang, "A deep learning based energy-efficient computational offloading method in Internet of vehicles," *China Communications*, vol. 16, no. 3, pp. 81–91, 2019.
34. X. Wang, X. Wei, and L. Wang, "A deep learning based energy-efficient computational offloading method in Internet of vehicles," *China Communications*, vol. 16, no. 3, pp. 81–91, 2019.
35. A. Gulati, G. S. Aujla, R. Chaudhary, N. Kumar, and M. S. Obaidat, "Deep learning-based content centric data dissemination scheme for Internet of Vehicles," Proceedings of the 2018 IEEE International Conference on Communications (ICC), Kansas City, MO, USA, 2018.
36. M. Liu, Y. Teng, F. R. Yu, V. C. Leung, and M. Song, "Deep reinforcement learning based performance optimization in blockchain-enabled Internet of Vehicle," Proceedings of the 2019 IEEE International Conference on Communications (ICC), Shanghai, China, 2019.
37. Y. Dai, D. Xu, Y. Lu, S. Maharjan, and Y. Zhang, "Deep reinforcement learning for edge caching and content delivery in Internet of Vehicles," Proceedings of the 2019 IEEE/CIC International Conference on Communications in China (ICCC), Changchun, China, 2019.
38. F. Kong, J. Li, B. Jiang, and H. Song, "Short-term traffic flow prediction in smart multimedia system for Internet of Vehicles based on deep belief network," *Future Generation Computer Systems*, vol. 93, pp. 460–472, 2019.

39. M. A. Hossain, R. M. Noor, K.-L. A. Yau, S. R. Azzuhri, M. R. Z'aba, and I. Ahmedy, "Comprehensive survey of machine learning approaches in cognitive radio-based vehicular Ad Hoc networks," *IEEE Access*, vol. 8, pp. 78054–78108, 2020. https://doi.org/10.1109/ACCESS.2020.2989870.
40. E. S. Ali, M. K. Hasan, R. Hassan et al., "Machine learning technologies for secure vehicular communication in Internet of Vehicles: Recent advances and applications," *Security and Communication Networks*, vol. 2021, p. 23, Article ID 8868355, 2021.
41. M. A. Alsheikh, D. Niyato, S. Lin, H.-P. Tan, and Z. Han, "Mobile big data analytics using deep learning and Apache spark," *IEEE Network*, vol. 30, no. 3, pp. 22–29, 2016.
42. D. M. West, *Moving Forward: Self-Driving Vehicles in China, Europe, Japan, Korea, and the United States*, Center for Technology Innovation at Brookings, Washington, DC, USA, 2016.
43. V. Mnih, A. P. Badia, M. Mirza et al., "Asynchronous methods for deep reinforcement learning," Proceedings of the 2016 International Conference on Machine Learning, New York, NY, USA, 2016.
44. H. Chiroma, S. M. Abdulhamid, I. A. T. Hashem, K. S. Adewole, A. E. Ezugwu, S. Abubakar, and L. Shuib, "Deep learning-based big data analytics for internet of vehicles: Taxonomy, challenges, and research directions," *Mathematical Problems in Engineering*, vol. 2021, pp. 1–20, 2021.

Section 4

Data Science Applications

Chapter 4

Minimized Channel Switching and Routing Protocol for Cognitive Radio–Based Internet of Vehicles

Muhammad Nadeem, Muhammad Maaz Rehan, and Ehsan Ullah Munir

4.1 INTRODUCTION

Internet of Things (IoT) and 5G technologies have opened up a plethora of opportunities for humans, resulting in increased connectivity and automation of daily tasks. Effective communication is crucial to enable this automation, and research indicates that there will be approximately 14 billion [1] devices by 2024. Additionally, there has been an average 40% increase in these devices, as reported in [2]. However, the sheer number of devices has led to a problem of spectrum scarcity, where it is impossible to assign dedicated wireless channels to each type of device.

According to the Federal Communications Commission, traditional devices only occupy between 15% and 85% of the available spectrum at any given time, leaving a significant portion of the spectrum idle. This underutilization of resources has led to the development of cognitive radio (CR) technology [3], which is an intelligent radio that can change its transmission and reception parameters. It is also known as software-defined radio (SDR) [4]. CR can detect spectrum holes and use them for communication, enabling the utilization of underutilized spectrum [5, 6].

CR-enabled networks consist of two types of devices: primary user (PU) and secondary user (SU). PU devices are allocated licensed channels by governing authorities such as the Federal Communication Commission (FCC) and the National Telecommunications and Information Administration (NTIA) in the USA [7]. These devices always have higher priority than SU devices. While each country has its own governing authority, the spectrum ranges for different communications (e.g. VLF, LF, MF, and HF) are globally defined. PU devices have complete authority to use the assigned channels per their requirements. In contrast, SU devices are equipped with CR and are required to sense available channels. They can only use channels that are not currently in use by PU devices. CR technology has important capabilities, including cognition, reconfiguration, and learning. Cognition enables a CR device to understand its geographical and operational environment, which allows it to sense the available spectrum. Reconfiguration allows a CR to dynamically and autonomously decide, based on the results of its cognition,

DOI: 10.1201/9781003284871-8

how to change its transmission and reception parameters. Additionally, CR devices can learn from their experiences and continuously improve their performance.

CR networks can be divided into two categories: infrastructure based and infrastructureless. The latter is also known as a cognitive radio ad hoc network (CRAHN), which is an appropriate choice for a VANET. Vehicles equipped with CR technology in a VANET are called cognitive radio–enabled VANET (CR-VANET). VANETs that have access to pedestrian, RSU, and internet resources make up the Internet of Vehicles (IoV). The IoV can be equipped with CR, known as CR-IoV. Although there are many challenges associated with CR communication, channel assignment is a widely explored area because it is a fundamental function of CR. There are three approaches used for channel assignment: static, dynamic, and hybrid [8].

The static channel assignment technique is the simplest approach, where each vehicle interface is assigned a permanent channel at the beginning of the network [8]. Since there is no competition among SU devices for channels and vehicles cannot change their selected channels, this technique results in very low utilization of available channels.

In the dynamic approach, channel assignment is recomputed based on changes in the network condition, such as the addition and deletion of PU devices. This approach is more efficient than static channel assignment because it allows for better utilization of available channels [8]. However, it comes at a cost, as the process of recomputing the channel assignment requires additional computational resources and may introduce delays in the communication process.

The hybrid approach for channel assignment is only suitable for cognitive radio systems that have multiple interfaces. In this approach, certain interfaces are allocated fixed channels, while others use a dynamic method to select channels [8].

4.1.1 Cognitive Radio Architecture

Figure 4.1 depicts an overview of CR. CR works on physical and link layers.

The physical interface is fulfilled by the radio, which serves as a medium for communication. The spectrum sensing module leverages this interface to detect and sense the availability of channels. CR technology can be implemented with a single or multiple physical interfaces. In a single-interface system, there is only one physical interface, whereas multi-interface systems consist of multiple physical interfaces. Each interface in a multi-interface system has its own spectrum control module at the physical and link layer [9].

The spectrum sensing module is responsible for detecting available channels and controlling the transmission and reception parameters. Algorithms are used to determine which channel selection method is appropriate for the given situation. Image contents are reused from [10].

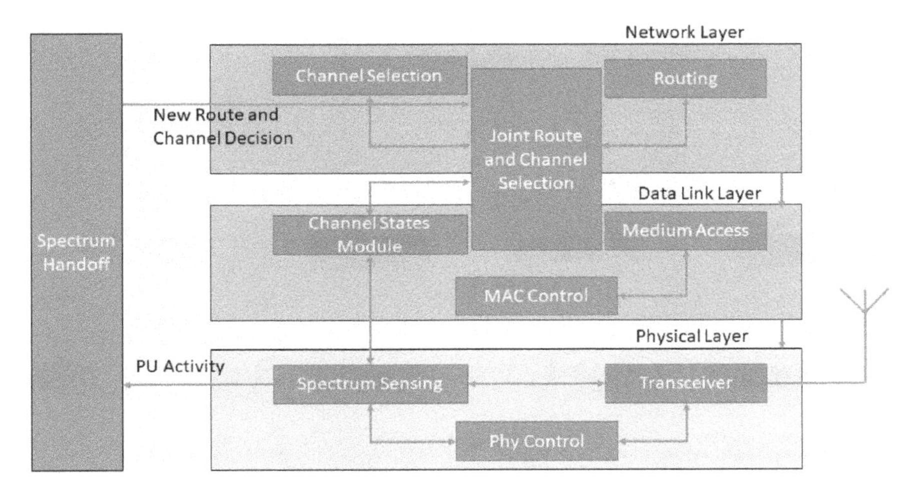

Figure 4.1 Architectural view of cognitive radio stack.

4.1.2 Cognitive Radio in the IoV

As previously mentioned, CR can also be integrated into the IoV, creating what is called CR-IoV. These networks are more flexible than traditional VANETs, as they are not limited to specific allocated channels and perform better in denser network scenarios. They also have access to roadside units and internet resources. Figure 4.2 illustrates a CR-IoV. Two approaches [11] are used for channel selection in CR-IoV, vehicle-to-vehicle (V2V) and vehicle-to-infrastructure (V2I).

The V2V approach involves vehicles exchanging control information with one another to determine the best available channel. A common control channel (CCC) is utilized for sharing control information among the vehicles. In contrast, the V2I approach utilizes roadside units placed along the road. Each vehicle shares its information with the RSUs and queries them for channel assignment. The RSUs function as the central entity for channel assignment within a particular region, selecting the best available channel in response to vehicle requests.

4.1.3 Channel Assignment in Cognitive Radio

Channel assignment algorithms have specific objectives [12] such as minimizing interference, maximizing bandwidth, reducing end-to-end delay, ensuring maximum throughput, and providing guaranteed packet delivery. The algorithms are designed to meet their respective objectives. Applications that require immediate response and assured packet delivery focus on achieving minimal end-to-end delay. The end-to-end delay is also affected by the channel switching delay, as frequent channel switching results in

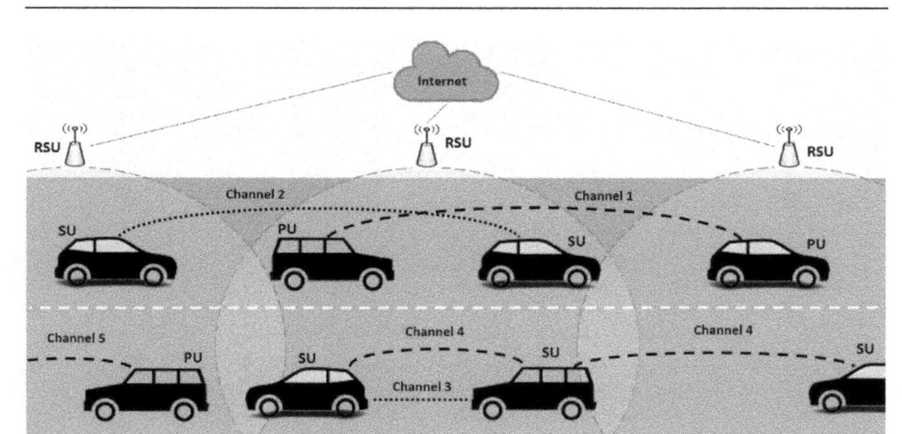

Figure 4.2 Cognitive radio–enabled Internet of Vehicles.

increased delay. This chapter primarily emphasizes the channel switching delay in the CR-IoV. The impact of channel switching delay can be demonstrated through an example. As the CR-IoV faces significant challenges in channel assignment, the delay in channel switching becomes a critical factor.

In Figure 4.3, PUs and SUs are distributed randomly, and a total of five channels $c1$, $c2$, $c3$, $c4$, and $c5$, are available between any two devices. There are two paths between source (S) and destination (D). On the first path, $P_1 = S \rightarrow SU_1 \rightarrow SU_2 \rightarrow SU_3 \rightarrow D$, only those channels are available between SUs which are not occupied by PUs. For example, $c1$, $c2$, and $c3$ are available between S and SU_1 because PUi has occupied $c4$ and $c5$. Similarly, $c4$ and $c5$ are available between SU_1 and SU_2 because PUo has occupied $c2$ and $c3$, and $c1$ is occupied by PUn. This applies to the rest of path P_1 and to the other path P_2.

4.2 LITERATURE REVIEW

This section delves into the prior research conducted in the area of cognitive radio. One of the key research problems in cognitive radio networks (CRN) is channel assignment. The issue of channel assignment will be examined for both stationary networks and mobile networks such as cognitive radio–enabled IoV.

There are two methods for implementing channel assignment: full spectrum knowledge and local spectrum knowledge [**13**]. Proposed channel assignment algorithms can be categorized into three approaches: (a) centralized, (b) distributed, and (c) decentralized [**8**].

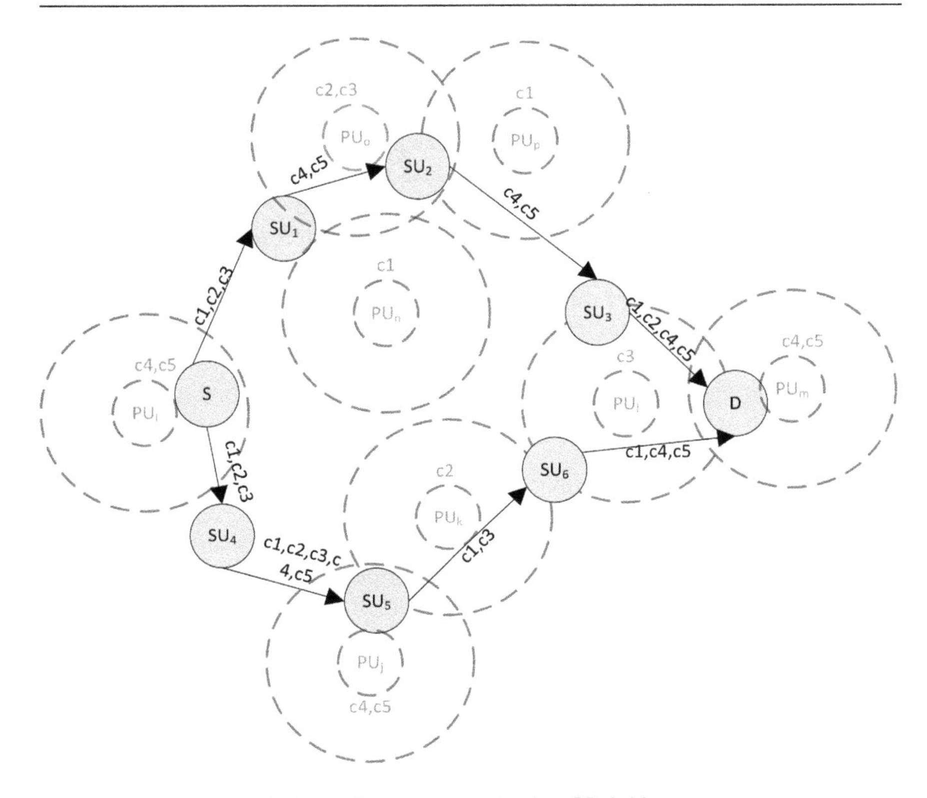

Figure 4.3 Example of channel assignment in the CR-IoV.

A distributed path selection scheme is proposed by [**14**], which creates clusters based on available spectrum, node power, and speed. The scheme employs both reactive and proactive routing, using proactive routing for intra-cluster communication and reactive routing for inter-cluster communication. Cluster heads (CHs) are elected based on the highest cluster head election value (CHEV), which takes into account the number of neighbors, number of common channels, and node speed. The stability of the cluster is maintained by prioritizing nodes with low speed as CHs. CHs periodically disseminate topology information to their cluster members, who maintain intra-cluster routing tables. For inter-cluster communication, a reactive approach is used with the help of the CH. If the destination is in the same cluster, the CH informs the source about the destination route. Otherwise, the CH broadcasts the message to adjacent clusters until the destination is found. However, this scheme is not suitable for networks with frequent changes.

The authors of [**15**] proposed an on-demand hybrid routing protocol for a cognitive radio ad-hoc network (CRAHN), which utilizes a cluster-based approach with neighbor quantity and available channels as parameters. To address the issue of frequent clustering, secondary cluster heads (SCHs) are defined by the cluster head. The protocol employs a super frame for information sharing, which includes a beacon period, spectrum sensing period, neighbor discovery period, and contention free period. During the beacon period, cluster ID, node location, node speed, and SCH information are shared. The accessible channel list (ACL) is updated during the spectrum sensing period, and nodes arrange the ACL in order and share information with their neighbors during the neighbor discovery phase. The contention free period is divided into mini-slots for intra-cluster and inter-cluster communication, with reserved mini-slots for handling new cluster members. Although mini-slot reservation for cluster members reduces channel conflict, it may result in wasted slots if a node has no data for intra- or inter-cluster communication.

The authors in [**16**] have suggested a scheme that uses OFDM to jointly allocate subcarriers and power for reliable packet delivery, with SINR being employed for detecting interference. Clusters of vehicles are formed, each with a common channel and a specific number of base stations. To eliminate co-channel interference, a transmission threshold value is determined based on the sender/receiver location, the transmission power of primary users, and the noise density of other primary and secondary users. OFDM enables low latency and jitter, facilitating real-time communication in IoVs. However, the technique proposed for channel selection is computationally expensive and memory intensive, which leads to higher costs.

A scheme for channel assignment is suggested in [**17**] that is controlled by topology. Both centralized and distributed algorithms are proposed for topology control. In the centralized approach, first the co-existing links of primary users are discovered; then the topology is constructed, and finally channels are assigned. In the first phase, nodes with minimum interference are identified, and the topology is formed using the Floyd–Warshall algorithm. For the channel assignment phase, two-hop conflict graphs are computed. On the other hand, the distributed algorithm consists of four stages: neighbor discovery using hello messages, topology construction using a minimum spanning tree and Steiner tree, power adjustment with one-hop neighbors, and channel assignment based on the resultant topology. The proposed scheme is capable of minimizing interference between secondary users and primary users and performs well during parallel transmission of both. However, the computational cost significantly increases with an increase in the number of nodes.

The authors discusses a matrix-based approach for channel assignment in [**18**]. The channel availability matrix and link interface matrix are used to obtain information about available channels and interference between links, respectively. The selection of channels considers important constraints, such as interface, interference, and channel availability. For meeting QoS requirements, a path with high capacity is selected. For real-time traffic, the number of hops is used as a benchmark. The data related to different requirements is piggybacked using Dijkstra's algorithm. However, this technique is suitable only for small network sizes, as the size of the matrix increases exponentially with the increase in the network size.

In [19], the authors propose a deep Q-learning–based approach that takes into account the available spectrum, the state of vehicular caching, and node mobility for making optimal decisions. The problem is modeled as a Markov decision process (MDP) where the set of vehicles is defined as the states of the MDP. The state transition of the MDP between two consecutive time slots is a combination of vehicle position and caching state. The MDP is solved using the Q-learning approach, which uses a Q-table to store learned state-action values. To achieve even better results, the Q-learning approach is combined with deep learning techniques. However, the computational cost of the proposed approach is high, which is addressed by using a dedicated server. This causes unnecessary delays in the system.

In [20], the authors proposed a distributed channel assignment technique that involves the installation of roadside units. Vehicles send their status information to the RSUs, and the channel is assigned through mutual coordination. The activity patterns of primary users are detected using a hidden Markov model (HMM) [21]. In this model, the hidden variables represent the PU activity, and the observed variables are the spectrum sensed by the RSUs. A positive weight is assigned to channels that have successfully transmitted a frame. This is a simple technique that does not involve extensive computation, but it is not suitable for dense networks. Additionally, for high-speed vehicles, this technique fails.

In [22], a double threshold (DTH) spectrum sensing technique is proposed for channel allocation. The authors proposed three threshold methods to reduce sensing overhead. The number of vehicles sending sense data to the fusion center (FC) is reduced through cooperative spectrum sensing (CSS). In CSS, both roadside units and secondary users cooperate with each other to find the probability of primary user activity. Time slots are defined for RSU-to-SU and SU-to-RSU data transfer. The RSU calculates a value taking different environmental factors into account and shares this with SUs in its range, which in turn use this value along with the detection value to decide whether they should report. Upper and lower threshold values are defined for SUs in DTH. The SU only reports sensing data when the value passes these thresholds. This cooperation increases the probability of making a correct decision. However, a higher number of environmental factors can interrupt synchronization due to computation delay, as the time duration is fixed for both RSUs and SUs for computation.

The authors of [23] have proposed a machine learning-based spectrum sensing technique for channel selection, which is a cluster-based hybrid approach with adaptive congestion-aware modeling. The proposed technique consists of four main components: a primary user base station (PBS), hybrid learning spectrum agents (HLSAs), SUs, and RSUs. A number of HLSAs work under the PBS, and each HLSA consists of a number of RSUs. HLSAs make intelligent decisions per the request from SU, as it periodically observes PU activity using a deep learning–based spectrum sensing algorithm. This algorithm uses learning for the signal environment, vehicle behavior, and network environment for optimal decision making. The proposed technique has shown excellent results, as it also predicts decisions due to deep learning. However, the implementation cost is high for this proposed work.

Authors in [24] proposes a decentralized approach for channel assignment in which roadside units receive messages from both primary users and secondary users to perform the assignment. This approach uses a two-stage detection technology to determine whether a channel is occupied by a PU or an SU. If the channel is occupied by a SU, other SUs can contend for the channel using any of the existing multiple access control (MAC) schemes. Each SU develops a spectrum sharing log, which is a 3D binary matrix that cooperates with other SUs to provide spectrum opportunities for the SUs' current and future positions, improving channel selection decisions. However, as the network density increases, the matrix size also grows rapidly, leading to delays.

The authors in [25] proposed a centralized dynamic spectrum sharing approach with privacy protection for PUs through geo-indistinguishability. Geo-indistinguishability is a technique to protect the location privacy of entities such as primary users in dynamic spectrum sharing scenarios. The available spectrum information is shared with a fusion center by the PU, which includes their position and channel in use. The FC entertains the request of secondary users and assigns the best available channel. However, if the FC's data is compromised, a malicious SU can interrupt PU performance. To ensure geo-indistinguishability of PUs, the authors proposed a dummy PU with a dummy location. The new PU is in the range of the original PU and shares valid channel information with the FC. This technique ensures location hiding of PUs. However, due to the central decision-making, this technique has a single point of failure.

The paper in [26] presents a study on radar-assisted cognitive radio–enabled vehicles, where the SU periodically senses signals using the Swerling 0, 2, and 4 models. The frame is divided into two slots: the radar sensing slot and the data transmission slot. A low-range radar is used in the first slot to detect the activities of the PU and SU. If both PU and SU are not within range, the radar receives noise, and if either PU or SU is present, the radar receives a signal that is the superposition of the radar echo signal and noise. Statistical properties of the radar cross section (RCS) are analyzed using the Swerling models. Swerling 0 deals with stationary vehicles, while Swerling 2 and 4 deal with temporal variations in RCS. The authors evaluate the proposed scheme using three performance metrics: detection probability, false alarm probability, and network throughput. However, the proposed solution underperforms, as several important parameters, such as vehicle mobility direction and current position of the vehicle, are ignored.

In the work described in [27], the authors proposed a cluster-based co-operative sensing approach aided by multiple input–multiple output (MIMO) antennas. In this approach, multiple antennas are used by each vehicle to sense the available spectrum, and the results are combined to detect the status of the primary user. The resulting data is then sent to a cluster head, which calculates statistics using the equal gain combining (EGC) and maximum ratio combining (MRC) techniques. The statistics are shared with a corresponding fusion center, which applies a K-out-f-N fusion rule to make a global decision in the case of stationary vehicles. For mobile vehicles, velocity is also considered for the global decision. The use of MIMO antennas led to a 13% improvement for the MRC rule and a 10% improvement for the EGC rule compared to a single input–single output (SISO) approach. However, the use of multiple antennas increased the overall cost of the network and resulted in a significant delay due to the decision-making hierarchy.

In [28], the authors proposed a spectrum allocation technique for CR-enabled vehicles that considers low and high loads. To target high throughput for low-load vehicles, the CASGA greedy algorithm is proposed. Meanwhile, to target maximum throughput for high-load vehicles, a separate scheme is used. In this technique, a single channel can provide multiple services simultaneously, each represented by a unique sequence with calculated transmission times based on the response threshold of the service and the transmission rate of the channel. The transmission rate is influenced by many factors, such as bandwidth, transmission power, Gaussian noise power, and the instantaneous gain of the vehicle. The proposed scheme aims to increase spectrum utilization with an increase in the number of vehicles.

4.3 PROBLEM STATEMENT

The literature review explores various techniques for channel selection, namely centralized, distributed, and cluster-based techniques. Centralized techniques involve sharing a significant amount of data for decision making, which can be memory intensive and impractical for large networks [17, 18, 22, 25]. These techniques lead to decreased performance and slow processing times. Some approaches have integrated machine learning and deep learning techniques, as seen in [19, 23, 27]. However, these AI-based techniques are computationally extensive and may not meet real-time requirements.

On the other hand, distributed techniques are well suited for environments such as CR-IoV, where extensive data sharing is not suitable, and vehicles make decision based on their local knowledge. These techniques, including [14, 20, 29], are fast and reliable and have no single point of failure. However, based on the literature review, several potential problems arise with channel selection in state-of-the-art distributed IoV-based cognitive radio techniques. These include high end-to-end delay, decreased throughput due to frequent channel switching at vehicles, and frequent channel reclaim by primary users.

4.4 RESEARCH QUESTIONS

Based on the problem statement and literature survey, following are the research questions which need to be answered.

1. How can the channel reclaim problem be efficiently resolved?
2. To achieve high throughput and the least end-to-end delay, how can channel switching be minimized in the CR-IoV?

The proposed minimized channel switching and routing protocol aims to resolve the aforementioned problems by selecting: (a) a stable channel and (b) the highest common channel, both on the end-to-end path.

4.5 PROPOSED MCSR

The proposed minimized channel switching and routing is a distributed V2V channel assignment scheme with a focus on optimizing data delivery between sender and receiver in a single-interface, multi-channel environment. In addition to data communication, a separate common control channel is utilized for sharing control information. Communication among vehicles is only possible when a suitable spectrum hole is available for transmission. The proposed MCSR system comprises three modules: (a) channel stats module (CSM), which maintains a record of channel usage by primary users and the load in terms of processed packets; (b) channel assignment module (CAM), which selects channels based on the CSM stats, the channel preference of the predecessor vehicle, and the currently occupied channels by one-hop neighbors; and (c) reactive routing module (RRM), which periodically sends MCSHello packets, broadcasts MCSRequest (CS-REQ) packets reactively, and receives MCSReply (CS-REP) packets. These modules work together to establish an efficient end-to-end path. When a vehicle needs to communicate with another vehicle, it transmits an CS-REQ packet. Figure 4.4 shows a structural view of the CS-REQ packet.

Figure 4.4(a) depicts three significant fields for the proposed MCSR protocol. The CS-REQ packet contains the top three channels available for communication. Three channels are modified by each intermediate vehicle to convey the best available channels to the next vehicle. The hop count value is incremented by one every time a CS-REQ is processed by a vehicle, indicating the number of hops between sender and receiver. The channel switch count field indicates the number of channel switches between the sender and receiver. If an intermediate vehicle chooses a different channel

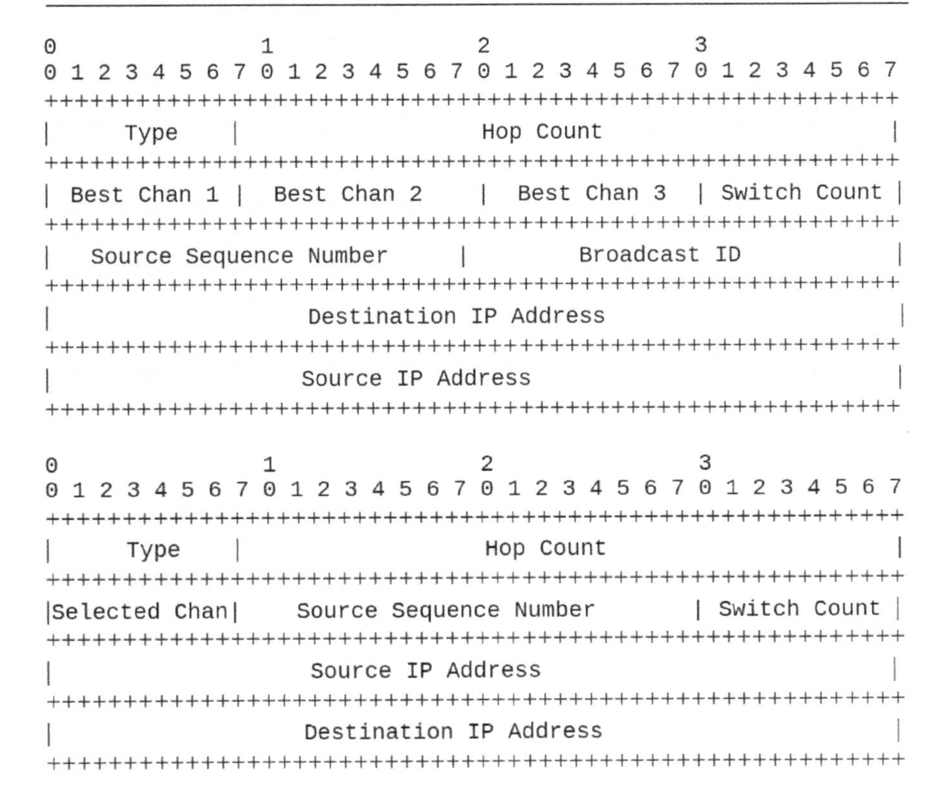

Figure 4.4 MCSR CS-REQ and CS-REP packet format.

than the previous vehicle, the count is incremented. Higher values of the channel switch count result in more end-to-end delay due to more channel switches. A path with a lower channel switch count has a higher likelihood of selection. The timestamp field is used by the source vehicle to determine the actual delay of a path and is only stamped at the source vehicle. Each vehicle uses this timestamp to identify end-to-end cost.

Upon reception of a CS-REQ, each vehicle executes Algorithm 1, which in turn uses Algorithm 2. A CS-REP packet is generated by the destination vehicle as soon it receives a CS-REQ. Figure 4(b) depicts the structure of this packet. A detailed explanation of the algorithm is given in next subsection.

Algorithm 1 Channel Selection Main Algorithm

```
Require: RREQ packet rreq for flow fl
 1: availableChannels[3] = 0
 2: if (v == rreq.src) then
 3:     rreq.availableChannels[]=bestAvailableChannels[SAC]
 4:     rreq.csc = 0
 5:     broadcast rreq
```

```
 6:     return
 7: end if
 8: if v=rreq.dst then
 9:     create a reverse route using rreq.availableChannels[0]
10:     generate rrep packet
11:     rrep.nextNodeChannel=availableChannels[0]
12:     unicast rrep
13:     return
14: end if
15: create reverse route channel, revrtc using rq.available
    Channels[0]
16: availableChannels[]=bestAvailableChannels[SAC]
17: if (revrtc∈availableChannels[]) then
18:     rreq.availableChannels[0]=revrtc
19:     rreq.availableChannels[1]=availableChannels[0]
20:     rreq.availableChannels[2]=availableChannels[1]
21: end if
22: broadcast rreq
23: return
```

4.5.1 Channel Assignment Algorithm

A CS-REQ is reactively broadcast by the source IoV after appending its top choices for channels. Algorithm 1 invokes Algorithm 2 to select the best available channels.

Algorithm 2 Best Available Channels

```
Require: Structured array of channels, SAC
 1: bestChannels[3]=0
 2: cr=0
 3: fl, flc
 4: pc, cpc
 5: hpc
 6: if totalchannel==2 then
 7:     bestChannels[0]=SAC[0]
 8:     return SAC
 9: end if
10: SAC[i].cr=0 for i=0, 1, 2,   C
```

11: $SAC[i].flc=\sum_{j=1}^{n} SAC[i].fl_j$

12: $SAC[i].cpc=\sum_{j=1}^{n}\sum_{k=1}^{m} SAC[i].pc_{j,k}$

13: $SAC[i].hpc=SAC[i].hpc+SAC[i].cpc$

14: sort ascending SAC based-on $SAC[i].flc$

15: $SAC[0].cr+=\omega_1$, $SAC[1].cr+=\omega_2$, $SAC[2].cr+=\omega_3$

16: sort ascending SAC based-on $SAC[i].cpc$

17: $SAC[0].cr+=\omega_1$, $SAC[1].cr+=\omega_2$, $SAC[2].cr+=\omega_3$

```
18: sort ascending SAC based-on SAC[i].hpc
19: SAC[0].cr+ = ω1, SAC[1].cr+ = ω2, SAC[2].cr+ = ω3
20: sort descending SAC based-on SAC[i].cr
21: bestChannels[0] = SAC[0], bestChannels[1] = SAC[1],
    bestChannels[2] = SAC[2]
22: return bestChannels
```

Algorithm 2 decides the best channel based on three parameters of available channels: number of active flows (*flc*), channel packet count (*cpc*), and historic packet count (*hpc*). This historic packet count plays a key role for channel history. The higher the value of the count, the busier the channel was in the past. This algorithm will return two channels if the CR-IoV is configured against only two channels (lines 6–8, Algorithm 2). If there are more than two channels, then Algorithm 2 further executes and computes the best available channels. First, all channel weights are set to zero (line 10, Algorithm 2). The receiver IoV of the CS-REQ consults its routing table to find all active flows against a channel and saves this count as *SAC*[*i*].*flc* (line 11, Algorithm 2). The number of packet transmitted on each channel is also saved in *SAC*[*i*].*cpc* (line 12, Algorithm 2). *cpc* answers with how busy a channel is. Channel busy stats can't be acquired through *flc*, as it does not provide any stats for traffic passing through a flow. The channel's history is stored in *SAC*[*i*].*hpc* (line 13, Algorithm 2). *hpc* helps to estimate the probability of channel availability in the near future.

Lines 14–19 (Algorithm 2) sort available channels based on *flc*, *cpc*, and *hpc* and assign corresponding weights to all parameters. These weights are accumulated in *cr* (lines 15–19, Algorithm 2). At the end, computed weights are sorted in descending order, and the channel with the highest *cr* value is selected as the best available channel 1; similarly, the second and third values correspond to the best available channels 2 and 3. These channels are returned to the main Algorithm 1 (lines 20–22, Algorithm 2).

The best three computed channels are returned by Algorithm 2 to Algorithm 1. The channel information is appended to the CS-REQ packet, and the source vehicle broadcasts this CS-REQ.

The destination IoV executes lines 8–14 of Algorithm 1 and unicasts the CS-REP packet. Each IoV receiving a CS-REP consults its routing table to find the channel selected as the reverse route and appends this information to the CS-REP packet. Forward routes are established by utilizing the channel information in the CS-REP packet.

When a CS-REQ is received by an intermediate vehicle, it establishes a reverse route using the best channel computed by the previous vehicle (line 15, Algorithm 1). Each intermediate IoV also computes its best available channel by using Algorithm 2. The intermediate IoV takes the intersection of the channel being used in the reverse route with newly computed channels. If the intersection is not null, it rearranges the best channel in such a way that the channel being used in the reverse route is assigned the highest priority (lines 17–21 Algorithm 1). This step plays a vital role in minimizing end-to-end channel switching. If the intersection is null, the IoV does not perform any rearrangement. After appending the channel information by the intermediate vehicle, the CS-REQ is broadcast so that it can be received by the ultimate destination vehicle.

4.5.2 Case Study

We can understand the proposed scheme with a case study. A vehicle entertaining a CS-REQ creates a reverse route, and a forward route is established using CS-REP. Both scenarios can be seen in Figure 4.5.

In Figure 4.5, there are seven secondary vehicles equipped with CR, and their transmission and interference ranges are shown using dotted circles. For simplicity, both the transmission and interference ranges are the same. The best channel computed by each vehicle is written alongside the vehicles, where channels in bold format are the actual channels transmitted by the vehicle. It is possible for the computed channels and transmitted channels of a vehicle to be different since the focus is on channel switching.

SU-1 is the source vehicle, and SU-5 is the destination vehicle in the given scenario. Algorithm 1 finds (1,3,4) as the best available channels, and this information is added to the CS-REQ. Only SU-3 is within the transmission range of SU-1. Upon receiving a CS-REQ from SU-1, SU-3 first establishes a reverse route with SU-1 using channel 1, which is the top-ranked channel from SU-1. SU-3 executes its algorithm and finds (3,5,6) as the best channels. The reverse route channel of SU-3 (i.e., channel 1) does not belong to any channel in (3,5,6), so SU-3 advertises the same channels (3,5,6) as its best available channels.

Channels advertised by SU-3 are received by both SU-1 and SU-6. SU-1 discards CS-REQ. SU-6 creates a reverse route using channel 3. Vehicle SU-6 finds channels (2,3,5) as its best available channel, but as a previous vehicle is connected using channel 3, it advertises channels (3,2,5) as the best available channels.

Channels disseminated by SU-6 are received by SU-4 and SU-3. SU-3 discards them, and SU-4 establishes a reverse route using channel 3. Channels (4,5,3) were found as the best available channels by SU-4, but channels (3,4,5) were advertised as its best channels. These advertised channels are received by SU-2, SU-6, and SU-5. SU-6 discards CS-REQ, while SU-2 establishes a reverse route using channel 3 and calculates (1,2,4) as the best channels. On the other hand, SU-5 was the destination vehicle, which replied with CS-REP.

When a CS-REQ packet is received by SU-5, it generates CS-REP by attaching the reverse route channel (3) to the CS-REP packet. This is required so that previous vehicles SU-4 to SU-5 can establish a forward route. SU-4 establishes a forward route using channel 3 and unicasts CS-REP to SU-6, attaching channel 3 with the CS-REP packet. SU-6 establishes

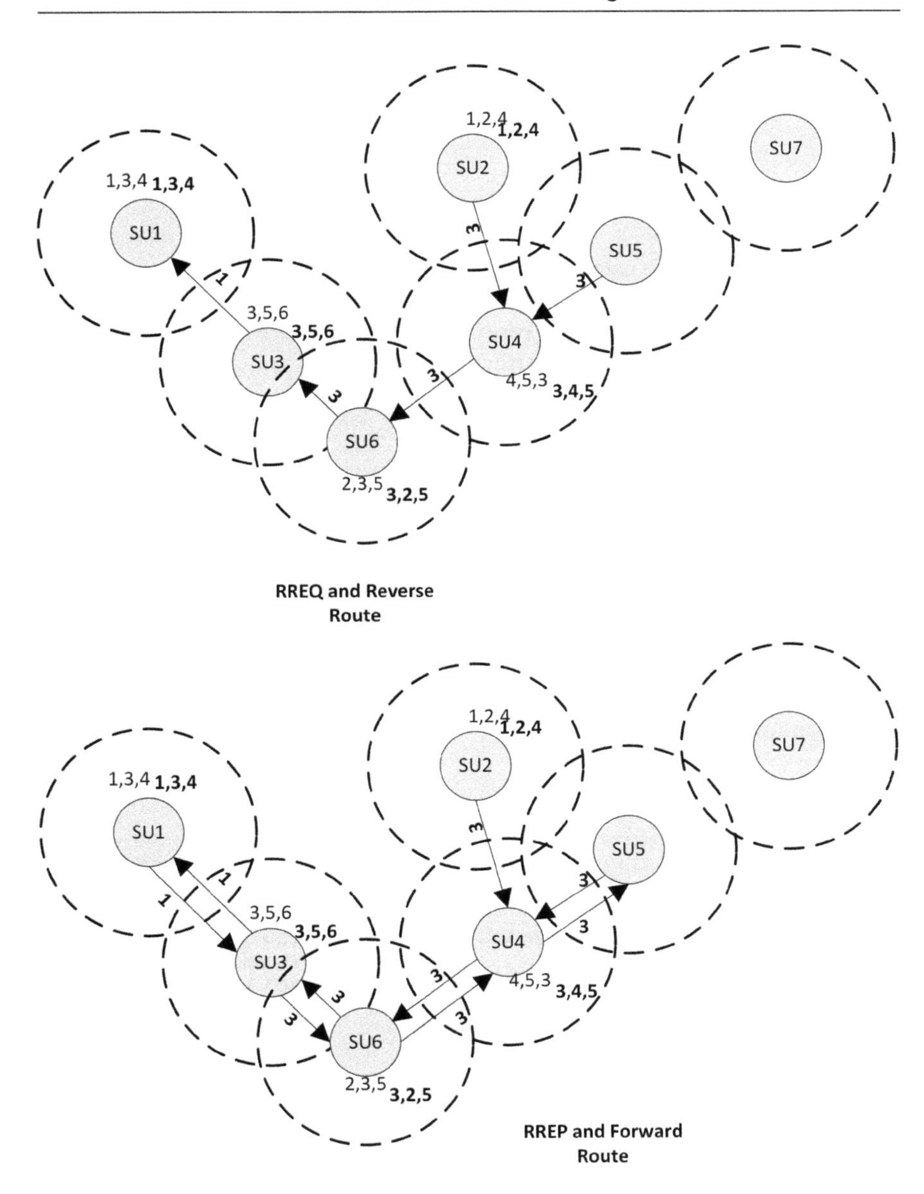

Figure 4.5 Case study, proposed MCSR protocol.

a forward route with SU-4 using channel 3 and unicasts CS-REP to SU-3, attaching channel 3. SU-3 establishes a forward route with SU-6 using channel 3 and unicasts RREP to SU-1, attaching channel 1. SU-1 establishes a forward route using channel 1. So a path with the best channels is established between source vehicle SU-1 and destination vehicle SU-5. The source and destination vehicle are three hops away, and using the proposed scheme, there is only one channel switched on the path. This considerably affects the overall end-to-end delay.

4.6 PERFORMANCE EVALUATION

This part briefly discusses simulation details and the obtained results. First, the simulation environment is discussed; then it is compared with the technique in [29], delay-minimized routing. The results show that our proposed scheme outperforms the competitor.

4.6.1 Simulation Environment

NS-2 is a widely used simulation tool for network communication and protocols, which was enhanced with the CRCN patch to enable simulation of cognitive radio networks. This patch can be downloaded from [30]. The simulations were conducted on Ubuntu 16.04 within a 7 × 7-Km simulation area using a real traffic SUMO map (Figure 4.6) of sector G9–1 Islamabad, Pakistan. The vehicles in the simulations were moving at speeds ranging from 60 to 70 km/h, and each simulation included three CBR connections. The final outcome was obtained by averaging the results of ten simulations.

Table 4.1 Simulation Parameters

Category	*Parameter*	*Value*
Physical Layer	Propagation model	Two ray ground
	Radio range CCA	250 m
	threshold	10 db
MAC Layer	Standard	802.11 g
	Interface queue type	DropTail/PriQueue 50
	Interface queue length	
Network Layer	Routing protocol	MCSR (proposed)
		DMR (competitor)
Transport Layer	Protocol	UDP
Application Layer	Protocol	CBR
	Packet size	512 bytes
Simulation Scenario	Area	7 × 7 Km
	Location	G9–1 Islamabad, Pakistan
	Number of vehicles	80
	Number of PUs	0,3,6,9
	Number of simulations	10
	Confidence interval	95%
	Network simulator	NS-2.31(CRCN patch)
	Time	300 sec

Figure 4.6 OpenStreetMap location—G9–1 Islamabad, Pakistan.

Sender and receiver were placed zero to three hops away in the simulation, and performance was measured against 0 primary user activity and 3, 6, and 9 PU activities. Table 4.1 depicts different simulation parameters.

4.6.2 Results and Discussion

The performance of both the proposed MCSR and DMR were evaluated against the achieved throughput and delay. Figure 4.7 gives a visual comparison of performance in terms of throughput for both protocols. Here sender and receiver were placed zero to three hops away, and performances were measured against 0, 3, 6, and 9 PU activities. A detailed analysis of the results indicates that the proposed MCSR exhibits superior performance compared to the DMR. When there is no activity from the primary user and no hop distance between the sender and receiver, the performance of both the proposed MCSR and DMR is nearly identical, as shown in Figure 4.7(a). However, as the number of hops increases, the DMR's performance declines in comparison to the proposed MCSR. When the distance is four hops away, the proposed MCSR achieves an almost 70% throughput, while the DMR only manages to attain close to 40%.

Observing the other graphs in Figures 4.7(b–d), it is apparent that the DMR experiences a decline in throughput performance as primary user activity increases. In contrast, the proposed MCSR exhibits satisfactory performance with an increase in primary user activity, as well as an increase in the distance between the sender and receiver.

A frequent channel reclaim by a primary user will result in higher packet drop and more end-to-end delay. This channel reclamation also severely affects throughput because the packet queue becomes overwhelmed, causing packet drops. Both of these issues are taken into consideration by the proposed MCSR in channel selection, as in Algorithm 1. The proposed MCSR prefers channels with less utilization by PUs in the past. A simple strategy is utilized: if a channel was less often reclaimed by PUs in the past, it more likely will also have lower chances in the future, so it will be a good choice to use this channel. The proposed MCSR not only considers history but also considers the total number of channel switches on the path between the source and destination. If we have a flow between two vehicles which are four hops away, and no channel is switched across the path, then it will have considerably higher performance as compared to the same path with three channel switches. This performance will be higher both in terms of end-to-end delay and throughput. This claim is supported by the results in Figures 4.7 and 4.8.

On the other hand, the DMR [29] takes into consideration the actual path delay for selecting a path, which undoubtedly enhances performance. But only considering delay for path selection is not enough, as other parameters also severely affect path performance. One factor is the total number of channel switches on the path. Another important factor is nonconsideration of channel behavior in the past in terms of PU activity. It is evident from the results that an algorithm not considering these factors faces a decline in performance.

Figure 4.8 compares both techniques in terms of the observed delay. The performance of the proposed MCSR improves with an increase in PU activity, whereas the performance of the DMR becomes worse with an increase in both the distance between the source and destination and PU activity. Another factor that can be observed is that the performance of the proposed MCSR remains consistent throughout all simulations. However, the DMR performs poorly when there are nine PU activities and the source and destination are four hops away [Figure 4.8(d)], leading to an exponential increase in delay. DMR performance will be severely affected if PU activities are increased; on the other hand, the proposed MCSR will have good performance with an increase in PU activities.

The results indicate that channel switching is a crucial factor in determining the achieved throughput and delay. Frequent channel reclamation by PUs leads to an increase in channel switches, which not only results in a higher end-to-end delay but also affects the achieved throughput.

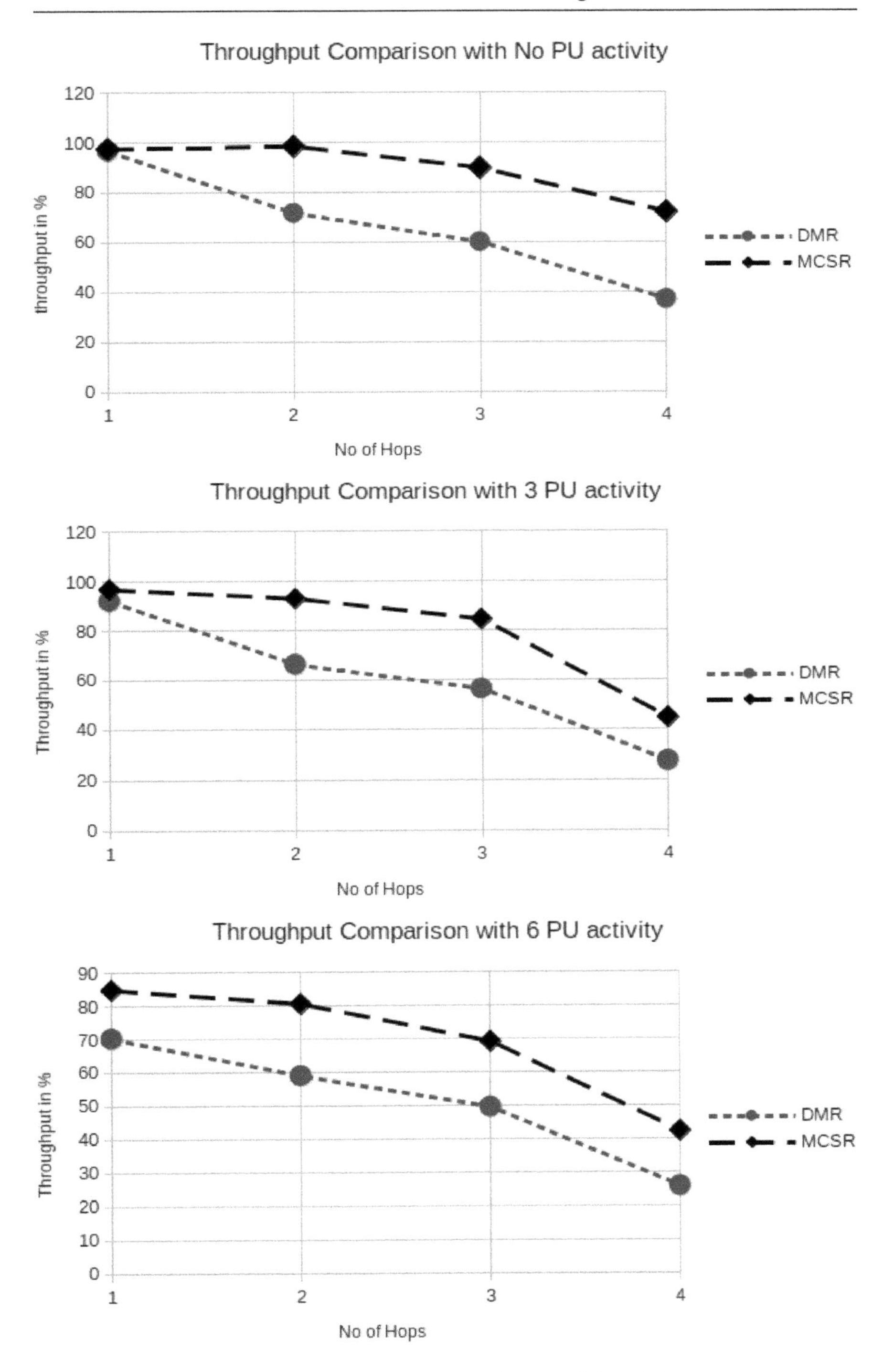

Figure 4.7 Throughput comparison against different PU activities.

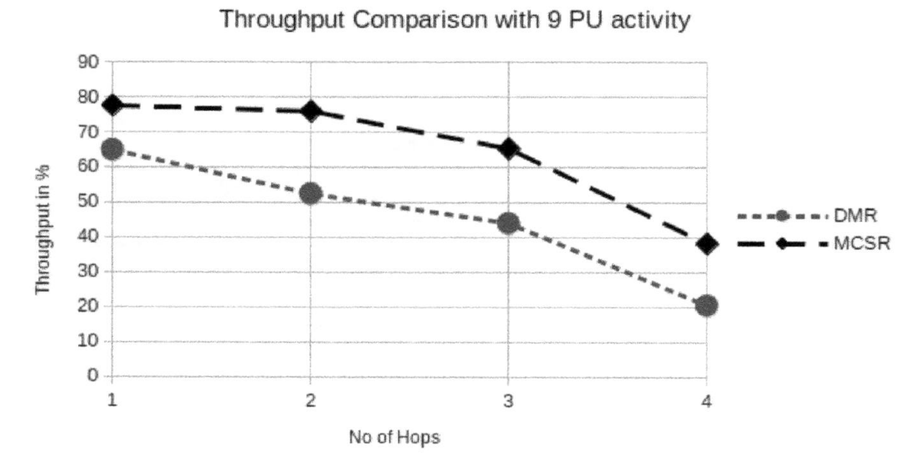

Figure 4.7 Continued

Figure 4.9 gives a summarized overview of the results. In Figures 4.9(a,b), the throughputs of the DMR and proposed MCSR are compared. The DMR is severely affected by the increase in hops between the source and destination and more primary user activities. On the other hand, the proposed MCSR attained almost double the performance of the DMR, while Figures 4.9(c,d) summarize both in terms of delay. These results show that in DMR, end-to-end delay significantly increases with an increase in number of hops between source and destination and PU activity. On the other hand, the proposed MCSR performance became better with an increase in PU activity and number of hops.

4.6.3 Conclusion and Future Work

This chapter has investigated the impact of frequent channel reclaim by PUs and more channel switches in the CR-IoV. It has been observed through simulation that the frequent channel reclaim problem in the CR-IoV leads to higher end-to-end delay and less throughput. A high number of channel switches between the source and destination further contributes to delay. Throughput and delay problems increase with an increase in PU activity and hop distance between source and destination.

To deal with this problem, a channel assignment protocol, MCSR, is proposed. The proposed MCSR assigns suitable channels to the CR-IoV by considering the channel history and total number of channel switches on a path. The frequent channel reclaim problem is dealt with by having a complete history of a channel, and a channel count field is used in the CS-REQ to deal with the higher channel switch count problem.

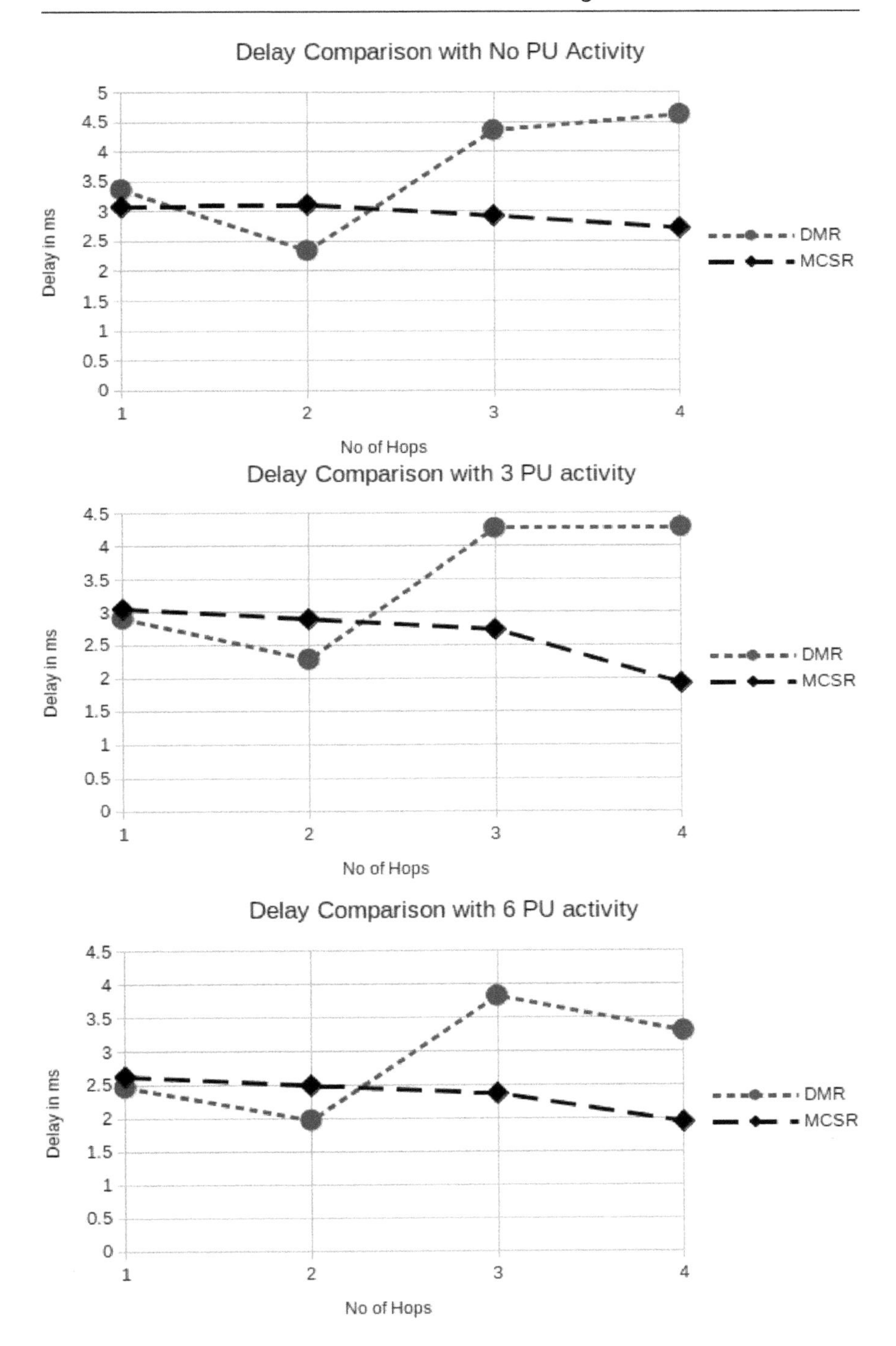

Figure 4.8 Delay comparison against different PU activities.

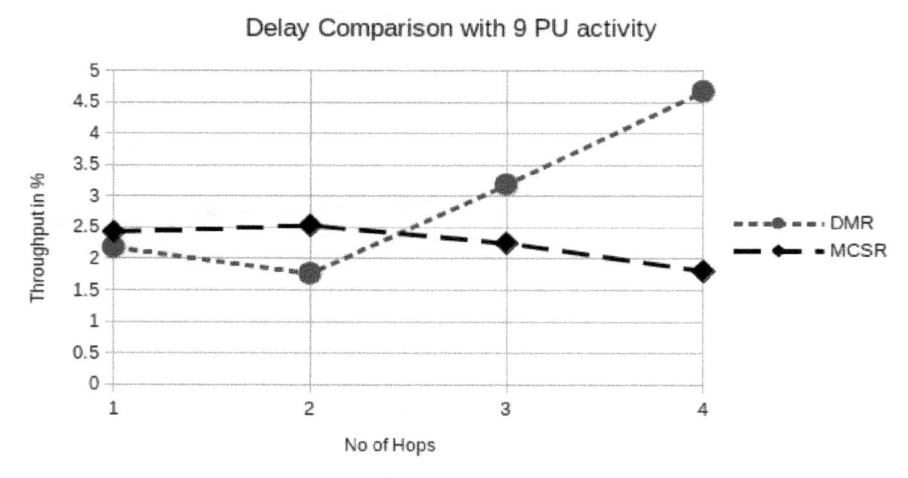

Figure 4.8 Continued

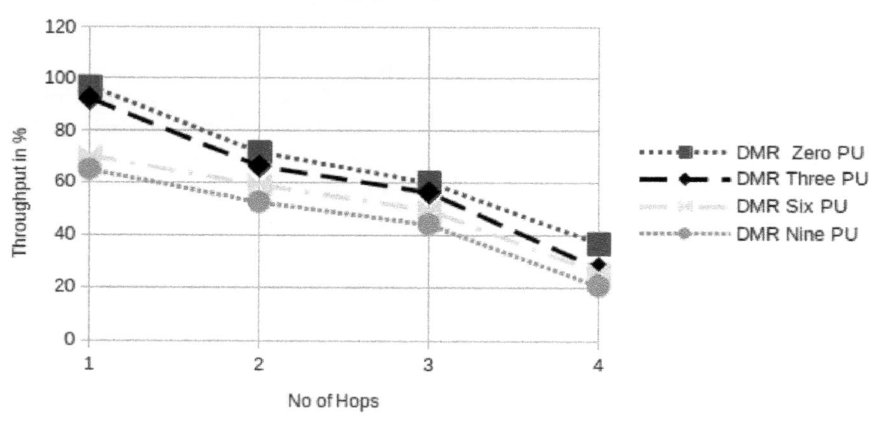

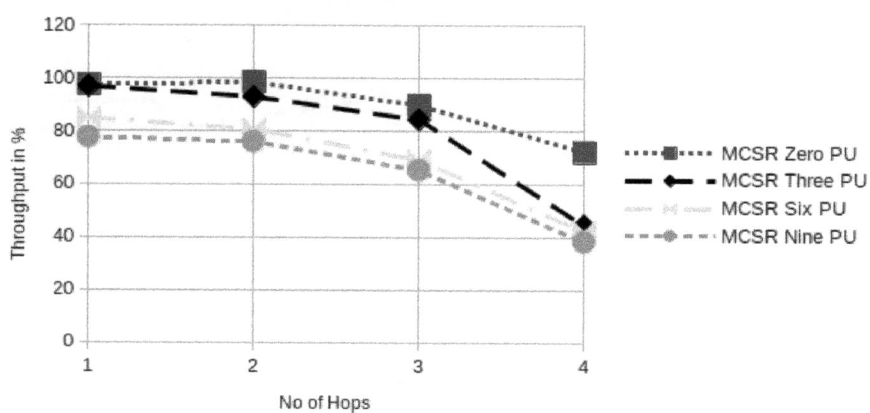

Figure 4.9 Summarized result for throughput and delay between DMR and MCSR.

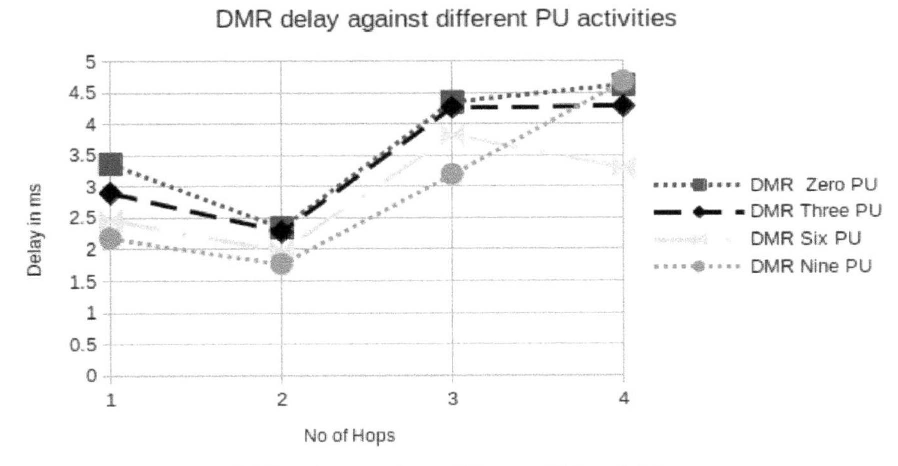

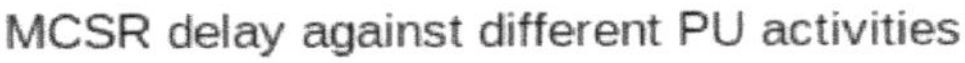

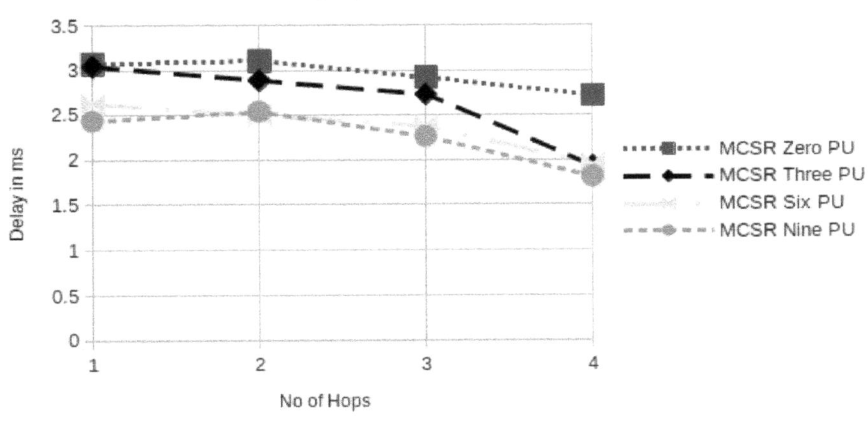

Figure 4.9 Continued

The proposed MCSR keeps track of active flows, packet count, and historic packet count of a channel. A source vehicle initiates a CS-REQ packet when desired for a path to a destination. Information on three channels is appended to the CS-REQ by the source vehicle by computing its best available channels. The intermediate vehicles create a reverse route with the best channel and computes its top three channels. The computed best three channels are compared with the channels of the previous vehicle; if the intersection is null, the computed channels are broadcast, and the channel switch count is incremented. If the intersection is not null, newly computed channels are rearranged in a such way that the highest priority is given to the channel which has been used in the reverse route. This rearrangement ensures minimum end-to-end channel switches. A destination vehicle unicasts the CS-REP packet, and all intermediate vehicles find channels in the routing table and append this channel information to the packet so that a forward route channel can be configured. The proposed MCSR is compared with the DMR. An extensive simulation was performed on a real-world map, and the results are compared in terms of throughput and delay. The results show that the proposed MCSR outperforms the DMR in terms of delay and throughput.

In the future, we have a plan to enhance the proposed MCSR by setting some benchmarks for channel assignment so that it can be used by vehicles with specific throughput and delay requirements. We are also planning to work on local channel recovery in case of PU appearance.

REFERENCES

1. Ramanpreet Kaur, Kavita Taneja, and Harmunish Taneja. Evolutionary computation techniques for strengthening performance of commercial MANETs in society 5.0. In *Evolutionary Computation with Intelligent Systems*, pages 181–198. CRC Press, 2022.
2. B Sanou. Telecommunication development bureau, ICT data and statistics division the world in 2013: ICT facts and figures, 2013. Accessed: 2024-04-16. https://www.itu.int/en/ITU-D/Statistics/Documents/facts/ICTFactsFigures2013-e.pdf
3. Ian F Akyildiz, Won-Yeol Lee, Mehmet C Vuran, and Shantidev Mohanty. Next generation/dynamic spectrum access/cognitive radio wireless networks: A survey. *Computer Networks*, 50(13):2127–2159, 2006.
4. Jacques Palicot. Cognitive radio: An enabling technology for the green radio communications concept. In *Proceedings of the 2009 International Conference on Wireless Communications and Mobile Computing: Connecting the World Wirelessly*, pages 489–494. 2009.
5. Paul Kolodzy and Interference Avoidance. Spectrum policy task force. *Federal Communications Commission, Washington, DC, Report ET Docket*, 40(4):147–158, 2002.
6. Jing Yang. *Spatial Channel Characterization for Cognitive Radios*. 2004.
7. Beibei Wang and KJ Ray Liu. Advances in cognitive radio networks: A survey. *IEEE Journal of Selected Topics in Signal Processing*, 5(1):5–23, 2010.
8. Ejaz Ahmed, Abdullah Gani, Saeid Abolfazli, Liu Jie Yao, and Samee U Khan. Channel assignment algorithms in cognitive radio networks: Taxonomy, open issues, and challenges. *IEEE Communications Surveys & Tutorials*, 18(1):795–823, 2014.
9. Thomas W Rondeau and Charles W Bostian. Cognitive techniques: Physical and link layers. In *Cognitive Radio Technology*, pages 219–268. Elsevier, 2006.
10. Sahil Vij. Cross-layer design in cognitive radio networks: Issues and possible solutions. *Network*, 7(8):9, 2019.
11. Joy Eze, Sijing Zhang, Enjie Liu, and Elias Eze. Cognitive radio technology assisted vehicular ad-hoc networks (VANETs): Current status, challenges, and research trends. In *2017 23rd International Conference on Automation and Computing (ICAC)*, pages 1–6. IEEE, 2017.
12. Alexander M Wyglinski, Maziar Nekovee, and Thomas Hou. *Cognitive Radio Communications and Networks: Principles and Practice*. Academic Press, 2009.

13. Matteo Cesana, Francesca Cuomo, and Eylem Ekici. Routing in cognitive radio networks: Challenges and solutions. *Ad Hoc Networks*, 9(3):228–248, 2011.
14. Mahdi Zareei, Ehab Mahmoud Mohamed, Mohammad Hossein Anisi, Cesar Vargas Rosales, Kazuya Tsukamoto, and Muhammad Khurram Khan. On-demand hybrid routing for cognitive radio ad-hoc network. *IEEE Access*, 4:8294–8302, 2016.
15. Nafees Mansoor, AKM. Muzahidul Islam, Mahdi Zareei, and Cesar Vargas-Rosales. RARE: A spectrum aware cross-layer MAC protocol for cognitive radio ad-hoc networks. *IEEE Access*, 6:22210–22227, 2018.
16. Joy Eze, Sijing Zhang, Enjie Liu, and Elias Eze. Design optimization of resource allocation in OFDMA-based cognitive radio-enabled Internet of Vehicles (IoVs). *Sensors*, 20(21):6402, 2020.
17. Min Sheng, Xuan Li, Xijun Wang, and Chao Xu. Topology control with successive interference cancellation in cognitive radio networks. *IEEE Transactions on Communications*, 65(1):37–48, 2016.
18. Zhihui Shu, Yi Qian, Yaoqing Yang, and Hamid Sharif. A cross-layer study for application-aware multi-hop cognitive radio networks. *Wireless Communications and Mobile Computing*, 16(5):607–619, 2016.
19. Ke Zhang, Supeng Leng, Xin Peng, Li Pan, Sabita Maharjan, and Yan Zhang. Artificial intelligence inspired transmission scheduling in cognitive vehicular communications and networks. *IEEE Internet of Things Journal*, 6(2):1987–1997, 2018.
20. Wei Yao, Abid Yahya, Fazlullah Khan, Zhiyuan Tan, Ateeq Ur Rehman, Joseph M Chuma, Mian Ahmad Jan, and Muhammad Babar. A secured and efficient communication scheme for decentralized cognitive radio-based Internet of Vehicles. *IEEE Access*, 7:160889–160900, 2019.
21. Brian G Leroux. Maximum-likelihood estimation for hidden Markov models. *Stochastic Processes and Their Applications*, 40(1):127–143, 1992.
22. Ellen Hill and Hongjian Sun. Double threshold spectrum sensing methods in spectrum-scarce vehicular communications. *IEEE Transactions on Industrial Informatics*, 14(9):4072–4080, 2018.
23. R Ahmed, Y Chen, B Hassan, L Du, T Hassan, and J Dias. Hybrid machine-learning-based spectrum sensing and allocation with adaptive congestion-aware modeling in CR-assisted IoV networks. *IEEE Internet of Things Journal*, 9(24):25100–25116, 2022.
24. Joy Eze, Sijing Zhang, Enjie Liu, and Elias Eze. Cognitive radio-enabled Internet of Vehicles: A cooperative spectrum sensing and allocation for vehicular communication. *IET Networks*, 7(4):190–199, 2018.
25. Xuewen Dong, Tao Zhang, Di Lu, Guangxia Li, Yulong Shen, and Jianfeng Ma. Preserving geo-indistinguishability of the primary user in dynamic spectrum sharing. *IEEE Transactions on Vehicular Technology*, 68(9):8881–8892, 2019.
26. Sai Huang, Nan Jiang, Yue Gao, Wenjun Xu, Zhiyong Feng, and Fusheng Zhu. Radar sensing-throughput tradeoff for radar assisted cognitive radio enabled vehicular ad-hoc networks. *IEEE Transactions on Vehicular Technology*, 69(7):7483–7492, 2020.
27. Mohammad Amzad Hossain, Michael Schukat, and Enda Barrett. Mumimo based cognitive radio in Internet of Vehicles (IoV) for enhanced spectrum sensing accuracy and sum rate. In *2021 IEEE 93rd Vehicular Technology Conference (VTC2021-Spring)*, pages 1–7. IEEE, 2021.

28. Ruifang Li and Pusheng Zhu. Spectrum allocation strategies based on QoS in cognitive vehicle networks. *IEEE Access*, 8:99922–99933, 2020.
29. Feilong Tang, Can Tang, Yanqin Yang, Laurence T Yang, Tong Zhou, Jie Li, and Minyi Guo. Delay-minimized routing in mobile cognitive networks for time-critical applications. *IEEE Transactions on Industrial Informatics*, 13(3):1398–1409, 2017.
30. CRCN Patch. https://kecsong.wordpress.com/2015/06/30/how-to-install-cognitive-radio-network-crn-in-ns-2/. Accessed: 2023–03–15.

Chapter 5

Time Synchronization in Cognitive Radio–Based Internet of Vehicles

S. M. Usman Hashmi and Muntazir Hussain

The cognitive radio–based Internet of Vehicles (CR-IoV) distributed network uses a new set of communication technologies in order to connect vehicles to vehicles, pedestrians, infrastructure, networks and clouds. CR-IoV nodes that are mobile vehicles, pedestrians and so on [1–4] are physically not connected with each other, which creates a wireless network of distributed and decentralized nodes. A simple representation of CR-IoV is shown in Figure 5.1.

These wireless networks improve traffic efficiency, provide better traffic management, enable efficient resource utilization, reduce accident rates and enable infotainment services [5–8]. Other applications of CR-IoV include road safety, reporting traffic jams and accidents and collision detection and avoidance [9]. The CR-IoV helps vehicles stay well connected to various transportation elements. Advanced technologies are implemented for CR-IoV systems to get the most out of them. With these advances in technology, vehicles can analyze more information and make it possible to build intelligent transport systems. It also encourages new revolutions in automated driving technology [10–12].

5.1 TIME SYNCHRONIZATION AND ITS IMPORTANCE

Distributed systems need a shared notion of causality to enable coordination. Due to interaction with physical processes governed by time-dependent laws, the causality notion is strengthened to establish a synchronized notion of time which can have microsecond precision requirements.

The CR-IoV is a distributed system whose applications are versatile in nature and have different requirements such as latency, reliability and throughput. The most challenging task is time synchronization among network nodes because of the decentralized and distributed nature of network, especially when nodes are mobile and moving on the roads [13–14]. CR-IoV systems have unique features, and their applications are time sensitive.

Event notifications are highly linked with time and location; hence nodes need to be synchronized to get the exact sequence of events. Event notifications without any order might not be useful in many scenarios. To report an

DOI: 10.1201/9781003284871-9

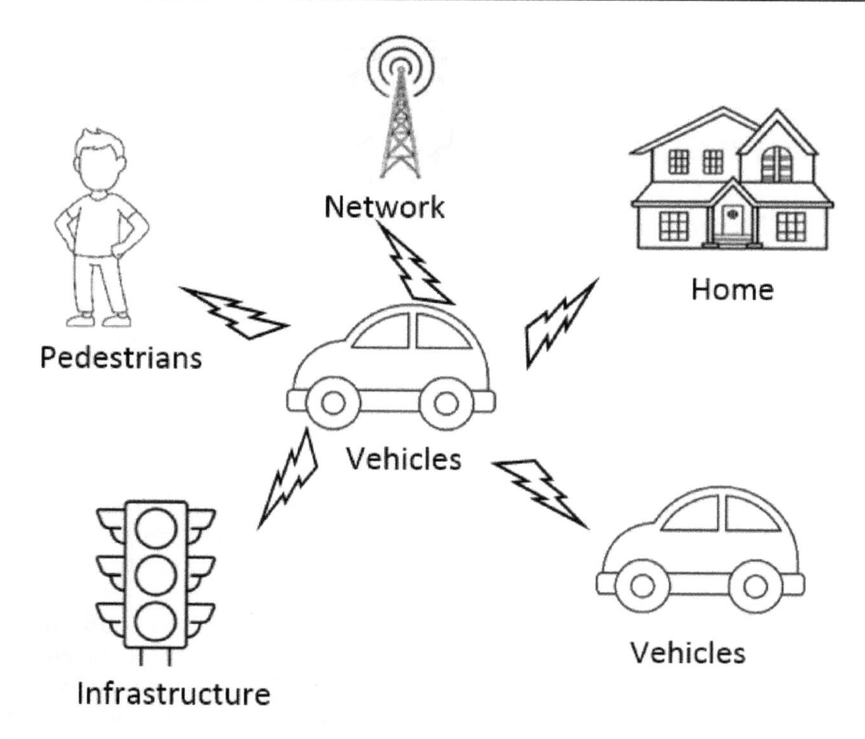

Figure 5.1 Cognitive-based Internet of Vehicles with distributed wireless nodes.

event at a specific time, each node must be synchronized in time; otherwise, each node will report a different time of the event [15–17]. For example, as shown in Figure 5.2, an accident took place, and its nearby nodes are out of sync and hence reported different times of the event. Most of CR-IoV applications are much less time offset tolerant: below 100 ms [18–20]. A single notion of time must be consistent throughout the network in order to facilitate the time-sensitive nature of CR-IoV services.

5.2 FUNDAMENTALS

Understanding time synchronization and the challenges in achieving it in distributed networks require the understanding of fundamental concepts. Let's first look at clocks and their workings and imperfections.

5.2.1 Clocks

A clock is a device that is used to measure time. Clocks indicate this measurement of time in a variety of ways. Analog clocks are based on mechanics,

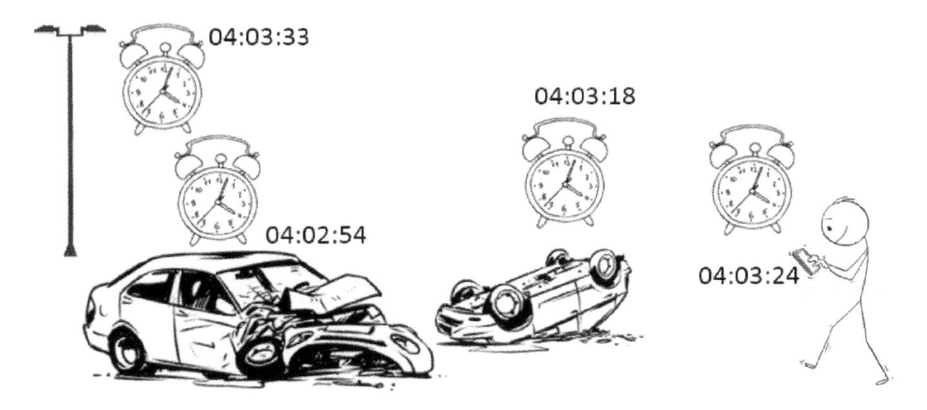

Figure 5.2 Nodes out of time synchronization.

and digital clocks use electronics to represent time. Clocks can be made for daily use and also for specific scientific purposes. They all keep time in their own manner. Most common clocks are mechanic, electric or atomic. In the CR-IoV, clocks can be either electric or atomic.

5.2.1.1 Electric Clocks

Every electronic device that has the capability to compute must keep track of time in order to process instructions in a sequential manner and to keep events in order. This time-keeping capability is provided by using a clock circuit. In a real sense, these clocks are actually timers. The heart of the clock circuit is quartz crystal, which is kept under tension [21]. Under this condition, quartz crystal oscillates, and the frequency of the oscillations produced depends on the type of crystal used and the provided tension. Each electric clock uses registers. The crystal oscillations become the input to the counter register and holding registers. These registers actually store integer values. A fixed value is stored in the holding register and is used as a reference to count oscillations before one clock tick. This value is first loaded in the counter register, and then on every oscillation of quartz crystal, the counter register decrements by one. When this counter register goes to zero, a clock tick is generated, and the counter register is reloaded from the holding register. This clock tick is actually a timer interrupt and used by digital processors and every other component of the computing device. Adjusting the crystal's frequency and value of the holding resister gives control of the frequency of generated clock ticks. Clock tick frequency is always less than the frequency of the crystal.

5.2.1.1.1 Actual Time Keeping

The clock ticks can be related to actual time in one of the following two methods. The first method has a specific date and time already stored in memory, and whenever such a system is turned on, it requires the current date and time, which can be taken from the user or any other source. The current date and time are converted into ticks from the start date that was already stored in memory and then stored in counting registers to continue from there. The second method is widely in use. Most computing devices have separate battery which keeps the clock running even when the device is powered off. In this way, devices can continue to boot up without asking any date–time information. This popular timekeeping approach is mostly used in clocks, watches, computers, electronic systems and communication networks like the CR-IoV [22].

5.2.1.1.2 Accuracy

The heart of the electric clock is quartz crystal, and that's the reason they are also known as quartz clocks. To keep the cost under control, manufacturers offer this clock with an error of a few seconds per week. If the cost is further required to be reduced, then a low-cost quartz crystal is used which may offer an error of one second per day, that is, around half a minute per month. Improving the quality of the quartz crystal can provide high accuracy, but it comes with a higher cost. Other factors that can affect accuracy are oscillator stability, which is used to count frequency oscillations, and variations in temperature.

5.2.1.2 Atomic Clocks

Another timekeeping approach is based on electron transition frequency in the electromagnetic spectrum of atoms. This type of timekeeping is the most accurate in existence. An atomic clock offers an error of a few seconds over trillions of years; hence it is the most accurate time and frequency standard known. However, accuracy comes with a high cost, which makes it unsuitable for computing devices.

Electric and atomic clocks both have pros and cons. Deploying atomic clocks in all nodes of a communication network is highly unreasonable due to the cost factor; using low-cost electric clocks in all nodes along with synchronization protocols seems to be a better choice.

5.2.2 Physical/Hardware Clocks

In order to design a typical node for the CR-IoV, some sort of actuator or transducer is used to sense and act accordingly. A transceiver for a communication and digital signal processor is also required, as shown in Figure 5.3. The processor performs every task as a sequence of instructions. It is important to provide this processor with some sort of timing information in order to maintain a sequence and execute one instruction at a time. Without a timing sense, all of the instructions will run simultaneously, and desirable output cannot be achieved. For this purpose, an electric clock, usually based on quartz

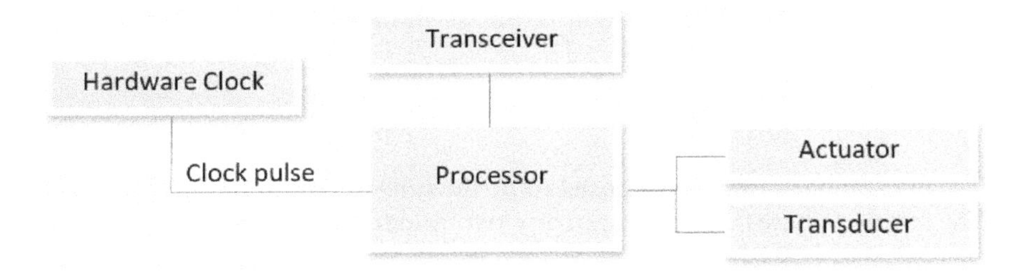

Figure 5.3 Structure of typical CR-IoV node.

crystal, is embedded with each processor. That clock ticks at regular intervals to generate a clock pulse and helps to synchronize the flow of instructions within the processor. This is known as a physical or hardware clock [23–25].

In order to generate a clock pulse, a hardware clock needs an oscillator and a counter. The oscillator generates a pulse train, and the counter is used to count and store the pulses. These clocks can be made from different materials, which decides their accuracy and precision, ranging from the most expensive caesium (atomic clocks) to low-cost quartz crystal (electric clocks).

5.2.3 Logical/Software Clocks

As the name suggests, a logical or software clock is a software-based programmable counter that uses counting algorithms to keep track of time. Logical clocks are derived from a node's hardware clock. Because of this dependency, the performance, precision and accuracy of the logical clock is based on the hardware clock. If $C(t)$ denotes the reading of a hardware clock counter t, then the software clock uses the value of $C(t)$ and converts it into time as $S(C(t))$.

5.2.4 Imperfections of Clocks

The heart of the software or hardware clock is the quartz crystal. Imperfections in the clock arise because of the quality of this quartz crystal and stability of the oscillator and counter. Stability depends on numerous factors, including initial frequency deviation, temperature, aging effect, jitter due to short-term noise and environmental changes. These factors deviate the clock from its actual time, and these clock uncertainties can be further described through phase offset, skew and drift.

5.2.4.1 Phase Offset

In any distributed network like the CR-IoV, each node of the network has its own clock [26–27]. Phase offset or phase error is the difference between the node's clock $C_A(t)$ and standard time t. Mathematically, phase offset for any node A can be written as,

$$\text{Phase offset} = \left|C_A(t) - t\right| \tag{1}$$

Usually standard time is supposed to be the time of the master node's clock. The relative phase offset between any two nodes A and B can be written as

$$\text{Phase offset} = \left|C_A(t) - C_B(t)\right| \tag{2}$$

Figure 5.4 shows two clocks with phase error. Clock A and Clock B are running at a phase difference of 20, and the difference remains constant for the given time t.

5.2.4.2 Skew

Clocks ideally must generate clock ticks of fixed length. The clock oscillator updates the counter at a constant rate or frequency, but due to the imperfections of the clock, the frequency may vary with time. This change in frequency causes the clock ticks to either slow down or speed up with respect to a perfect clock, as shown in Figure 5.5.

The difference in clock frequencies between the node's clock and the perfect clock is known as skew. Mathematically, clock skew can be computed

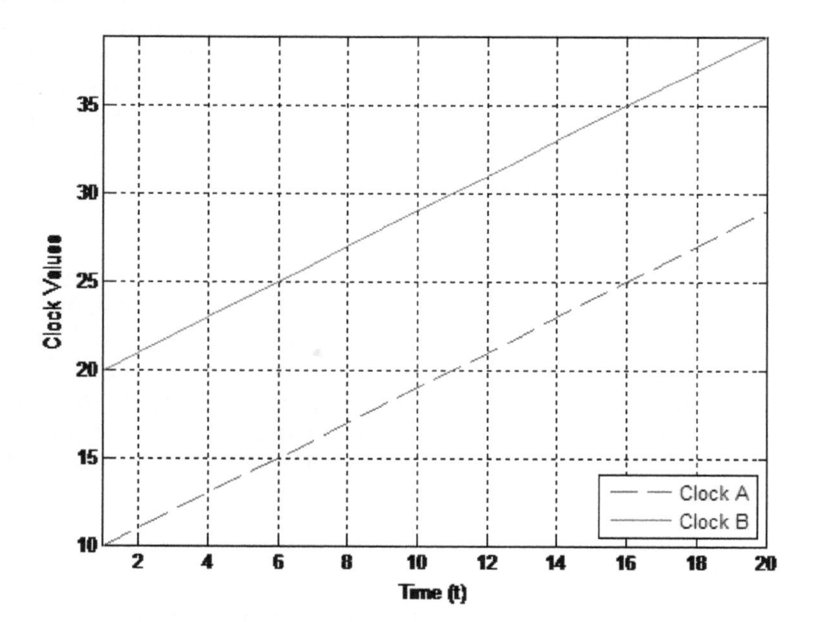

Figure 5.4 Clock phase offset between two nodes.

by taking the derivative of the phase offset with respect to standard time $C'(t)$ [28–29]. In order to compute the skew between two nodes A and B, the derivative of equation (2) can be written as,

$$\text{Skew} = C'_A(t) - C'_B(t) \quad (3)$$

Figure 5.6 shows two clocks with skew offset. Clock A and Clock B are running with different speeds. Initially the clocks are at zero phase offset, but with the passage of time, phase offset is introduced because of the different frequencies of the clocks.

In the CR-IoV, any two nodes with their own clocks can have phase offset as well as skew offset [30–31]. Figure 5.7 shows two clocks running with different frequencies with initial phase offset.

Clock A and Clock B are running with different speeds, and initially the clocks have a phase offset of 20. With the passage of time, phase offset is increased, and the clock values keep diverging from each other.

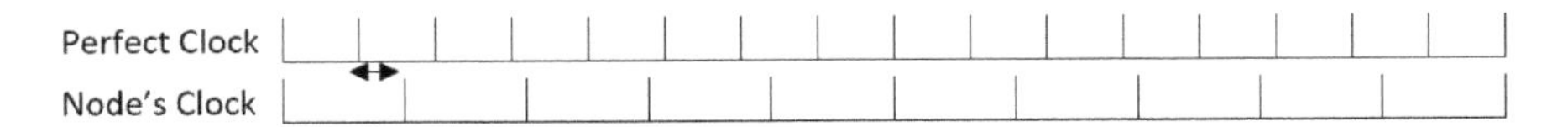

Figure 5.5 Variation in frequency and clock ticks.

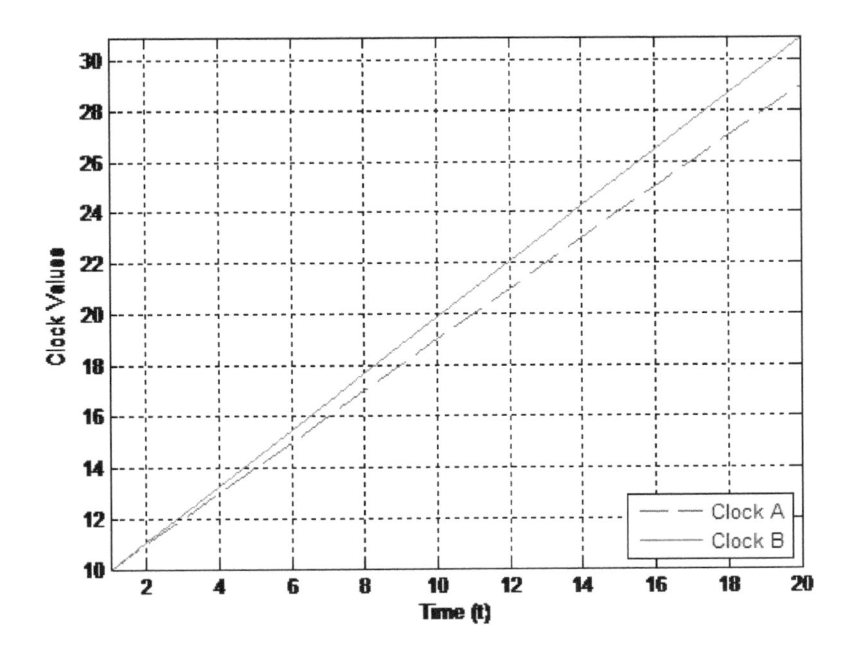

Figure 5.6 Clock skew between two nodes.

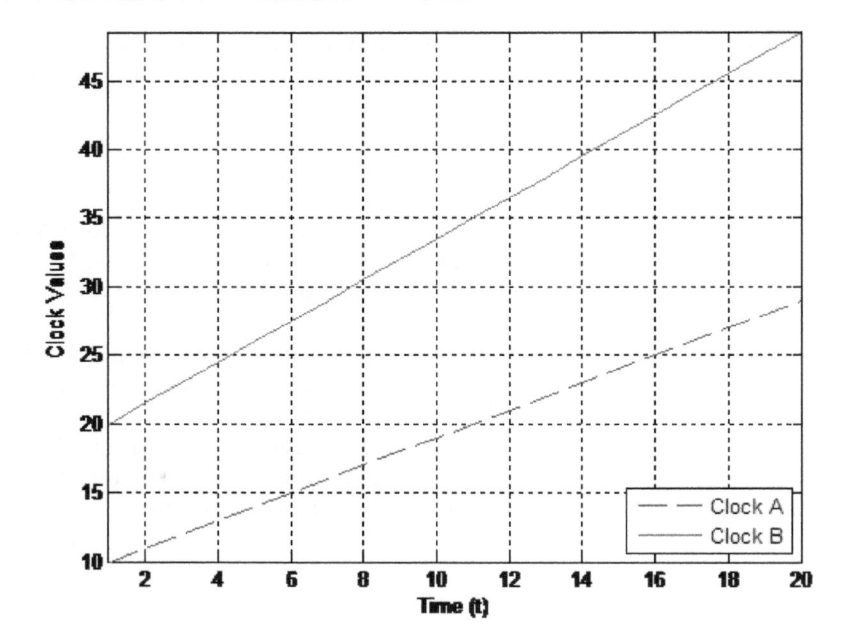

Figure 5.7 Clock skew between two nodes.

5.2.4.3 *Drift*

The drift of a clock can be obtained by taking the second derivative of the clock value with respect to time $C''(t)$. The drift between two clocks can be obtained by again taking the derivative of equation (3) and can be written as,

$$\text{Drift} = C_A'(t) - C_B'(t) \tag{4}$$

5.2.4.4 *Ideal Clocks*

An ideal clock is one that does not change its frequency and generates clock ticks of fixed length. These clocks do not deviate from real time. For an ideal clock *C(t)*, the rate of change at time *t* must be equal to 1 and can be written as,

$$\frac{dC(t)}{dt} = 1 \tag{5}$$

5.2.5 Synchronized Clocks

When clocks are running with the same frequency and zero phase offset, then these clocks are known as synchronized clocks. Figure 5.8 shows two synchronized clocks with no phase and frequency offset (skew).

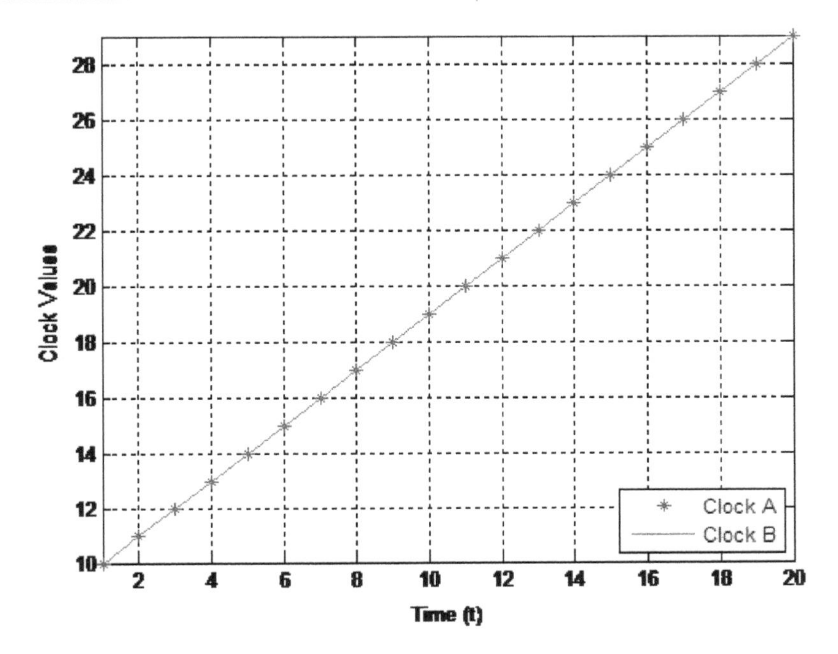

Figure 5.8 Synchronized clocks.

5.2.6 Clock Adjustments

Clock adjustment is a way of correcting clocks in such a way that it must not alter the working of the overall network. Two things can be done if clocks are found to be incorrect. Clock values may be corrected, or they may be allowed to run untethered [32].

5.2.6.1 Clock Correction

Clock values need to be updated to synchronize. This correction can be done by instantaneously changing the clock value or gradually varying the frequency.

5.2.6.1.1 Instantaneous Correction

Instantaneous correction can be done by assigning new values to the clock's counter register and adjusting the frequency. This method has its own setback. It can cause skipping of events. If some event is triggered within the time that is skipped, then the node will not be able to identify that event.

5.2.6.1.2 Continuous Correction

Continuous correction is way of adjusting clock frequency in order to match in phase gradually over a period of time. This method is more suitable because it avoids skipping of events. Change in frequency does the correction of the clock over time; hence no event can be skipped because the node has every clock value.

5.2.6.2 Untethered Clocks

Another way of keeping clocks synchronized is to save the offset in a register and let the clock run without any amendment in its value. Each node saves its offset, and whenever nodes need to communicate with each other, only then is the saved value used. In this way, no events are skipped, and clock adjustment can easily be done. Letting the clock run untethered is a popular method because of its advantages [33].

5.3 SYNCHRONIZATION CONCERNS

5.3.1 Global Time Access

The need for global time arises when all nodes desire a common notion of time. In general, there are two ways of achieving global time.

5.3.1.1 Installation of a GPS Device

A global positioning system (GPS) provides the time service known as GPS time. This GPS time is accurate up to around 14 ns. GPS devices can be installed in each node of the network, which will make every device capable of accessing GPS time. This eliminates the need of synchronization because whenever time is required by the node, it will be taken from a satellite. However, this method has its own limitations, like accessibility of satellites in every environment. This method will increase the cost of nodes, and eventually such a network of nodes will become too expensive to be practical.

5.3.1.2 Using Synchronization Protocols

A better approach is to install the GPS in a few nodes or just one master node. All of the remaining nodes must synchronize in time with the master node. This can be done by numerous methods, which we call synchronization protocols. Some of the most popular protocols are discussed later in this chapter.

5.3.2 Limitations

Establishing this shared notion of time remains a challenge. A clock is a computing device which is derived by the oscillator that produces a periodic signal at a known frequency. This idea of keeping time has two fundamental limitations. First, nodes initialized at different times will reflect phase offset in their clock values. Second, there will be variation in the frequency of the oscillators among different nodes, that is, frequency offset. When many clocks among a network calculate and compensate for the phase and frequency offset, then 100% efficiency can never be achieved.

5.3.3 Critical Metrics

Synchronization is the key element in any distributed network. The CR-IoV network uses time synchronization for coordination of nodes, sequencing and relating events with time. Various services offered by the CR-IoV have different requirements. Achieving the required level of precision and efficiency and keeping the cost low is the challenge. For time synchronization, critical metrics can be precision, efficiency, lifetime, cost and scope.

5.3.3.1 *Precision*

Precision of a clock refers to the maximum error that a clock can have in a network with respect to the master clock. The CR-IoV application decides what level of precision is required.

5.3.3.2 *Efficiency*

Synchronization algorithms or protocols require computations and message transmissions, and eventually energy will be consumed. Synchronizing in time must be energy and time efficient. Achieving this target of synchronization must be done with few computations and few message transmissions to save execution time as well as transmission time.

5.3.3.3 *Lifetime*

The lifetime of the synchronization refers to how often time synchronization becomes a requirement. A protocol that offers a longer lifetime means the number of synchronization rounds will be reduced. This will increase the network performance. However, it must be kept in mind that increasing the lifetime of synchronization improves energy efficiency but will drastically affect precision.

5.3.3.4 *Cost*

Cost is one of the most important factors and can be improved by making the design of nodes simpler and using energy-efficient protocols.

5.3.3.5 *Scope*

Synchronization must adapt to network size and change. Overall efficiency must not be affected by increasing the number of nodes.

5.3.4 Basic Principles

In any distributed network like the CR-IoV, the basic principle of time synchronization involves sharing of timestamps among nodes [34]. Let's consider a network with two nodes A and B that need to synchronize with each other. Node A will generate a message with its clock value, known as a timestamp. It will send this timestamp message to node B, and node B will update its clock value. This type of synchronization is unidirectional and will not compensate for message transmission delay. Accuracy is compromised in this way. In a distributed network, delay is a serious concern; hence protocols following the idea of unidirectional synchronization are not preferred. Another idea is bi-directional, or round-trip, synchronization, which provides promising results. In this approach, node A sends a timestamp at time t_X to node B, and node B will respond back by sending its own timestamp at t_Y. Node A will receive the response of node B at t_Z. The two-way transmission delay, known as round trip time, can now easily be computed as $d = t_Z - t_X$ and incorporated, as shown in Figure 5.9. A single timestamp exchange will only compensate for phase offset; however, multiple timestamps can be exchanged to estimate and compensate for skew. This will also increase message exchanges, affecting energy efficiency, but provides better accuracy. Most of the CR-IoV time synchronization protocols follow this principle.

5.3.5 Challenges

The basic principle of synchronization is based on sending and receiving timestamps. The synchronization technique has to face many challenges. Most importantly, it must be capable of dealing with transmission delays [35]. These transmission delays are random and can cause serious estimation

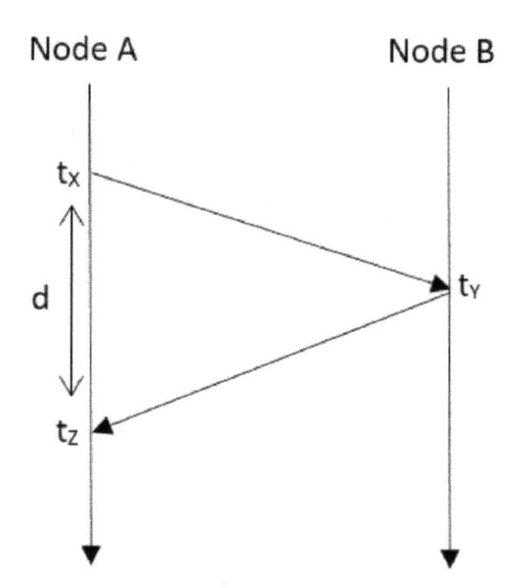

Figure 5.9 Basic principle of synchronization.

errors. Other latencies that can affect synchronization efficiency are sending time, access time, propagation time and receiving time.

Timestamping accuracy is another major issue. It is the method of adding and extracting time from the message. Once the node receives the correct time, the next phase is clock adjustment. The clock adjustment principle also determines the quality and lifetime of synchronization.

5.3.6 Various Types

Clock synchronization is related to many parameters, like source of real time, required accuracy and communication model. Because of its requirements in versatile applications, synchronization methods are also versatile. Some require internal consistency of time, and some require synchronization with real time.

In absolute clock synchronization, wireless systems like the CR-IoV maintain synchronization with a standard reference clock time. In this method, synchronization is implemented with respect to an external time standard. Timestamps are distributed among nodes by using any global radio system like satellite-based GPS. In relative clock synchronization, wireless nodes are time synchronized with respect to each other. In this method, estimation of phase offset and skew is made by exchanging timestamps within the network among nodes. Such synchronization with an internal timescale maintains a relative time with respect to each other.

5.4 APPROACHES TO TIME SYNCHRONIZATION IN WIRED MEDIA

Time synchronization is not only a problem of wireless networks but also of wired networks. Clocks of every node need to be synchronized so that they can relate events to time and time of day. The network time protocol (NTP) is the most widely used in wired networks for time synchronization. It creates a hierarchical tree of time servers and distributes the time information to the servers and then to the clients. The simple network time protocol (SNTP) is a subset of NTP and is the simple implementation of NTP. The precision time protocol (PTP) uses a master–slave architecture. The master node provides time information to the slave nodes and is synchronized with a grand master node attached to a time reference such as a global positioning system. In the time transmission protocol (TTP), a node sends its timestamp to the synchronizing node. The node that receives multiple timestamps estimates the time of the source node with the help of message delay statistics. Other protocols include Cristian's protocol and set-valued estimation. These protocols exchange multiple timestamps between a synchronized node and synchronizing node and estimate clock values using different algebraic formulas. None of the mentioned protocols are suitable for wireless environments. As energy is a very scarce resource in comparison with wired networks, time synchronization needed to be redefined for wireless networks.

5.5 APPROACHES TO TIME SYNCHRONIZATION IN WIRELESS MEDIA AND THE CR-IoV

Wired network time synchronization protocols do not offer very good results in wireless sensor networks because of the different requirements, that is, energy efficiency, infrastructure, end-to-end latency and reliability. Using GPS in every CR-IoV node to maintain the same time in the network is also an option, but it makes inexpensive nodes more expensive and energy consuming. Time synchronization in wireless networks can be done in three different ways. Relative timing seems to be the simplest one, as it only relies on the order of the events and does not maintain an actual clock, but it can only be used in limited types of CR-IoV applications. Another way of synchronizing is that each node has information about its frequency and phase offset with other nodes and can synchronize their clock values. This approach is mostly deployed by many time synchronization protocols. Global synchronization, another method of synchronization, tries to maintain a global clock throughout the network.

As a CR-IoV is a type of mobile ad-hoc wireless network, many protocols are available for time synchronization. A few of them are discussed in the following.

5.5.1 Protocols

Researchers have found several ways to compensate for both time phase and skew correction. Phase offset is calculated with message exchanges, with several novel mechanisms taking care of the uncertainties. Frequency offset can be eliminated by using stable but expensive oscillator. A more traditional and low-cost approach is to estimate the frequency offset using regression techniques on the results of several message exchanges. A significant amount of research in frequency estimation has gone into making it more accurate while using fewer message exchanges, thus conserving energy. The following protocols are designed specifically considering the requirements of wireless and ad-hoc environments like the CR-IoV offers. However, these protocols are more suitable for fixed or less mobile nodes like networks, infrastructure, slow-moving vehicles or pedestrians with a CR-IoV network.

5.5.1.1 Reference Broadcast Synchronization

Reference broadcast synchronization (RBS) uses beacons for time synchronization. RBS can be applied to single-hop or multi-hop networks. For a single-hop network, each node can send a beacon to broadcast the timestamp. The neighboring nodes receive the timestamp and compare it with their own clock values. It maintains a table that contains the local clock values of each node in the network. This table is used to relate the clocks of each node to each other and let the clocks run untethered. This leads to correction of the clocks of receiving nodes. RBS is capable of performing phase and skew correction in a whole network. For a multi-hop network, a group of nodes is made. These groups are known as clusters. One beacon is sent across one cluster for synchronization. A node, known as gateway node, is responsible from sharing of the timestamp from one cluster to another. Hence synchronization can be achieved throughout a multi-hop network. This synchronization can compensate for both phase offset and skew.

RBS can be used to decrease non-deterministic latency using receiver-to-receiver synchronization. Each node broadcasts beacons using a physical layer and computes the non-determinism of packet send time, access time and propagation time using the receive time. Figure 5.10 shows RBS deployed in the CR-IoV.

The process used by RBS for node synchronization for the estimation of time is shown in Figure 5.11.

5.5.1.2 Romer's Protocol

Romer's protocol uses an innovative time transformation algorithm. The lower bound and upper bound round trip time (RTT) are used to estimate message delay, which in turn is used for clock correction. It maintains a table that contains the local clock values of each node in the network. This table

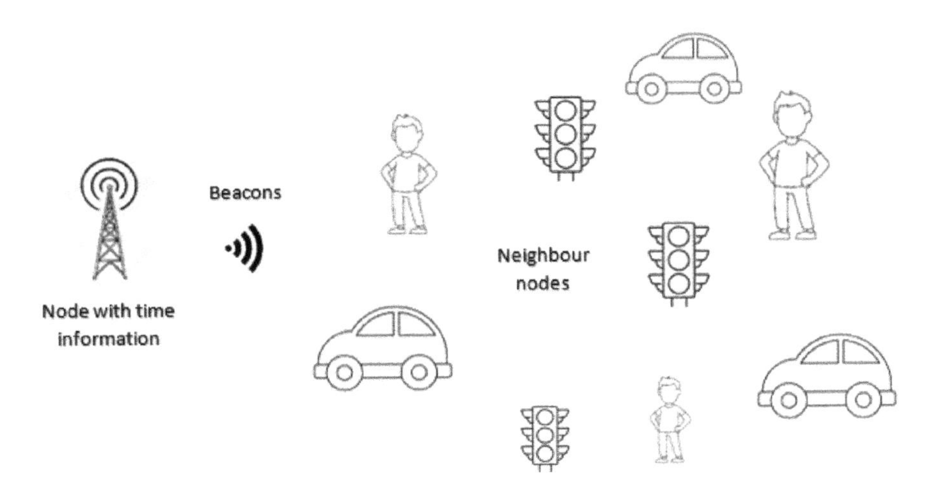

Figure 5.10 RBS deployed in the CR-IoV.

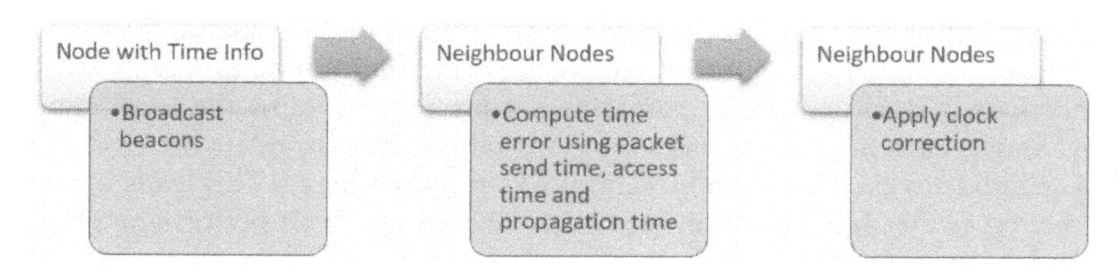

Figure 5.11 Process of RBS node synchronization.

is used to relate the clocks of each node to each other and let the clocks run untethered.

5.5.1.3 Timing-Sync Protocol for Sensor Networks

The timing-sync protocol for sensor networks (TPSN) is a sender-receiver–based synchronization method. This network-wide synchronization protocol is based on a hierarchical approach. A hierarchal topology is created that divides the nodes into multiple levels. Hence the synchronization task is performed in a hierarchal manner and in two phases. These phases are the level discovery phase and synchronization phase. For the level discovery phase, a master node with the true time is considered level 0, which is the root note. This root node sends a level discovery message to its immediate neighbors. The level discovery message that is broadcast includes the node's identity and level in the hierarchy. Nodes that receive this message assign themselves as level 1. Now level 1 nodes broadcast the level discovery message to their neighboring nodes. Similarly, receiving nodes assign themselves level 2 in the hierarchy. This process is repeated until all the nodes of the network are assigned a level number. The nodes are now ready for the synchronization phase. In this phase, all nodes synchronize their clocks to their parent node, which is higher in the hierarchy. It uses round-trip synchronization at the medium access control (MAC) layer; therefore, message delay uncertainties are mostly eliminated. This improves the accuracy of TPSN but at the cost of message exchange overhead. As the number of nodes increases, message overhead further increases. Figure 5.12 shows the hierarchy of TPSN.

Figure 5.13 shows the network synchronization process using TPSN. As shown in Figure 5.13, the level discovery phase consists of allocating a root node and assigning a level to each node, whereas the synchronization phase contains node synchronization with upper-level nodes and network synchronization to the root node.

5.5.1.4 Flooding Time-Synchronization Protocol

The flooding time-synchronization protocol (FTSP) is a promising hybrid protocol for time synchronization in wireless networks [36–37]. It is based on

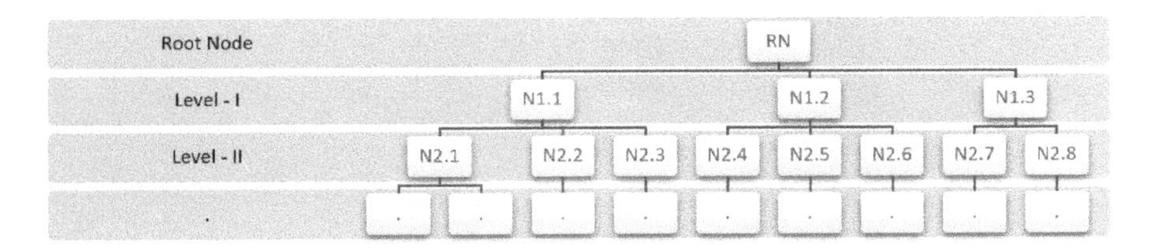

Figure 5.12 Hierarchy of TPSN.

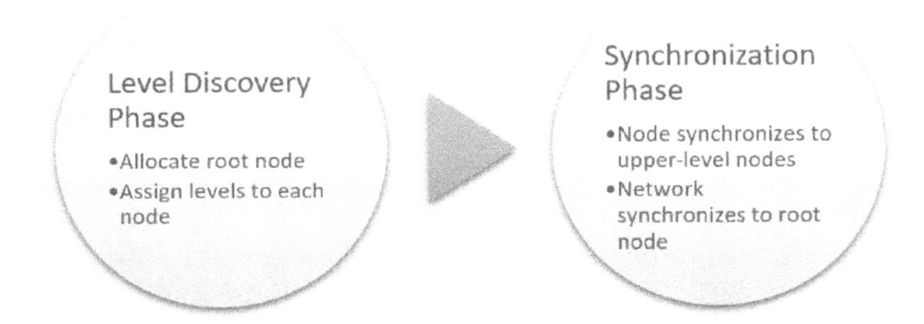

Figure 5.13 Process of TPSN node synchronization.

the principles of RBS and TPSN. FTSP follows a self-organized, fault-tolerant and bandwidth-efficient algorithm. The root node is the lowest-level node with a true clock which broadcasts synchronization messages within the network. All receiving nodes correct their clocks by comparing the timestamps. The root node can be replaced with its next-level node in case of failure. Using FTSP, a root node can synchronize every node in its range by using a single radio message timestamped at both the root and receiving nodes. Similar to TPSN, FTSP also uses MAC layer byte-wise timestamping and hence eliminates delays in radio message delivery. Figure 5.14 shows the process of FTSP node synchronization.

FTSP can work in single-hop as well as multi-hop networks. In an FTSP deployed network, each node is assigned a unique ID, root ID, sequence number, local clock, and regression table. A synchronization message contain a root ID, sequence number and timestamp. Following this method, FTSP can synchronize in nodes that are not in the direct range of the root node, that is, FTSP multi-hop time synchronization. Suppose node A has the true time and is selected as a root node, as shown in Figure 5.15. Root node A floods the time synchronization message and synchronizes with nodes B, C and D. Node D floods the time synchronization message and synchronizes with nodes E, F, G that are in broadcast range of node A. Similarly, node G synchronizes nodes H and I. By following this principle, multi-hop

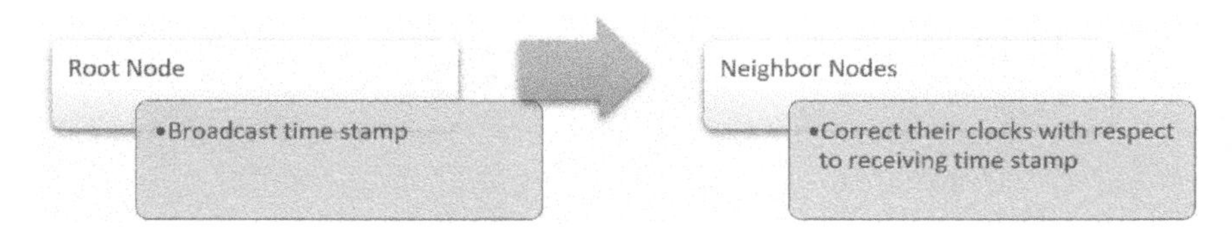

Figure 5.14 Process of FTSP node synchronization.

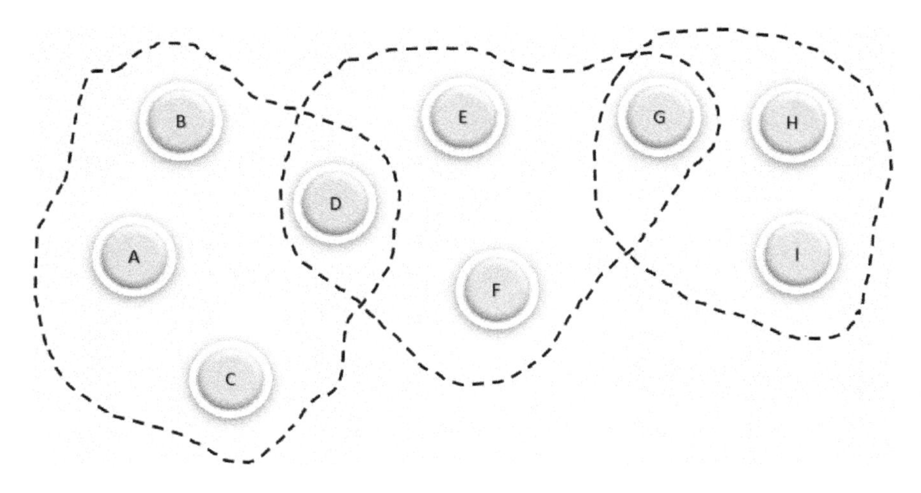

Figure 5.15 FTSP multi-hop synchronization.

synchronization is achieved. As FTSP is also robust, suppose if root node A fails, then node H can be selected as the root node based on a root node selection method. However, nodes become out of sync and resynchronization is needed in this scenario.

5.5.1.5 Time-Stamp Synchronization

The time-stamp synchronization (TSS) method is based on synchronization on demand. It implements internal clock synchronization. It does not generate any specific time synchronization packet separately. It adds the timestamp in other packets to perform synchronization. The rest of the protocol works the same and uses transmission delays to estimate the correct time. This protocol can be deployed in multi-hop networks. Synchronization in this way reduces the required number of message transmissions, hence conserving energy.

5.5.1.6 Lightweight Time Synchronization

Lightweight time synchronization (LTS) uses fewer message exchanges and provides a specific level of precision. It creates a spanning tree for the network and performs pair-wise synchronization. The root node of the tree is considered the master node with the true clock. It is responsible for sharing its clock value and also defining the lifetime of the synchronization. This protocol can be used in CR-IoV applications where precision is not a big problem because the average synchronization error using LTS is around 0.4 s, which is quite high in comparison to other protocols.

5.5.1.7 Cognitive Radio–Synchronization

Cognitive radio–synchronization (CR-Sync) is based on TPSN. A tree structure is created to assign levels to each node within the CR-IoV. Multiple timestamps are exchanged between parent and child nodes to synchronize the whole network. CR-Sync is a promising protocol for the CR-IoV because of its self-adaptive, fault-tolerant and mobility-based features. The major disadvantage is that it lacks robustness to root node failure.

5.5.1.8 Bio-Inspired Synchronization

Bio-inspired synchronization (BSynC) distributes nodes in the CR-IoV and categorizes them as master, ordinary, reference ordinary and neighbor nodes. Global navigation satellite system (GNSS) or a master clock is equipped in master node, and the whole network needs to be in sync with this. Ordinary nodes are those which are trying to synchronize. The reference ordinary node is an ordinary node nominated by another ordinary node to compute their phase and skew offset with each other in order to synchronize. Every node that falls under radio coverage of a node is considered its neighbor node. BSynC initiates the request to synchronization procedure (RSP) and time adjustment procedure (TAP) after node classification. Per RSP, the master node shares information with neighbor nodes regarding the available radio frequency channels, timestamp and master node ID. For the next procedure, TAP, the receiving nodes create the master node as a reference ordinary node and compare its clock value to the received timestamp. Eventually the node corrects its own clock. BSynC uses multiple timestamp exchanges in order to achieve phase and skew offset.

5.5.1.9 Miscellaneous Protocols

Many protocols are available in the literature for synchronization purposes. Each of them can be used in the CR-IoV depending on the application and its requirements. A few other protocols include the following: delay measurement time synchronization, which is an energy-efficient protocol but less accurate than the RBS protocol. Probabilistic clock synchronization extends RBS by providing probabilistic bounds on the accuracy of clock synchronization. The time-diffusion protocol (TDP) is based on diffusion of messages involving all the nodes in the synchronization process.

5.5.2 Significance in the CR-IoV

Time synchronization in any wireless distributed network system is a well-known issue. It is considered the key element for real time control and data operations.

Each node of any distributed system like the CR-IoV has its own clock. These clocks drift away in clock values with the passage of time. One of the causes of this phase offset is variation in clock ticking frequency. This poses a grave problem for time-sensitive CR-IoV applications. In the CR-IoV, networks have to relate events to time; hence each node has to associate a timestamp with the events happening. Time plays a vital role in the determination of two discrete real-world events. The time interval between two events may also be of interest. Even event ordering can't be done without time synchronization. Time synchronization protocols ensure that all distributed nodes, including vehicles, pedestrians and infrastructures, have a common notion of time so time-dependent issues can be resolved.

In many CR-IoV applications, true or actual time is very important. The requirements of such applications cannot be fulfilled by logical time or simply ordering events. They deploy time-based decisions which rely on actual time. One such service offered by the CR-IoV can be traffic management for each vehicle on the road. This can be done by sharing messages among nodes which include vehicle dynamics, speed, position, road information and driving intentions, along with real time of the day. Other examples of time-dependent CR-IoV applications to enable road safety can be forward collision warning (FCW), cooperative collision warning (CCW) and emergency electronic brake lights (EEBL). They may help the driver to avoid a possible accident. Each vehicle needs to periodically share the location and other data to generate warning messages. These warning messages are generated on highly time-sensitive information. If nodes among the network are not synchronized, then such information may report a past timestamp and advanced timestamp, and, in either case, the information is discarded or mixed up. The user of such an application, who is supposed to get a warning before an accident, may receive the warning after the accident or a false warning. Hence, the warning message would fail to alert drivers, thus leading to a risk of an accident. Time synchronization in the CR-IoV is therefore essential.

Time synchronization is also important for efficient bandwidth utilization and channel scheduling. Absolute time synchronization is also important to exercise security measures in the CR-IoV to prevent session hijacking and jamming.

5.5.3 Requirement Analysis for Time Synchronization in the CR-IoV

Having a common notion of time is a requirement of various CR-IoV applications. Synchronization principles ensure that all nodes have the same time in the wireless and distributed networks. Time synchronization in the CR-IoV needs a fine blend of accuracy, performance, feasibility, and compatibility. Requirements can be categorized into two classes: system-end requirements and application-end requirements.

5.5.3.1 System-End Requirements

System-end requirements include system issues, spectrum utilization, and compatibility of the system with the synchronization method. The wireless spectrum is a very scarce resource, and efficient utilization is one of the major requirements in any distributed wireless network. The CR-IoV has a limited spectrum. Optimized usage of the spectrum leads to enhanced bandwidth and throughput. This spectrum is divided into service and control channels. All wireless nodes must be synchronized for efficient channel coordination. However, wireless nodes go out of sync with time. A guard band is used as a guard interval to accommodate out-of-sync nodes. It is a period of time used to separate two consecutive transmissions which depends on synchronization accuracy. If CR-IoV nodes are efficiently synchronized, then the guard interval can be reduced. A reduced guard interval exponentially improves the overall spectral efficiency. For example, in IEEE 802.11n, reducing the guard interval to half leads to an 11% increase in the data transmission rate. Therefore, precise time synchronization enhances system performance.

Another system-end requirement is to ensure quality of service (QoS) in the continuously varying nature of the CR-IoV environment. Any CR-IoV node can join or leave a network, and this may cause a drastic change in node density. This scalability issue is better addressed if nodes are tightly synchronized. Another such issue is high mobility. As CR-IoV nodes are vehicles, they are highly mobile. Maintaining the same QoS in such an environment is again dependent on time synchronization.

5.5.3.2 Application-End Requirements

The CR-IoV environment and applications are highly dynamic. CR-IoV vehicular nodes can move as quickly as 150 Km/h or more. Such high speed means the connectivity of the vehicle node with its neighboring nodes will be for just few seconds, and it keeps changing neighbors. Moreover, some CR-IoV applications require very low end-to-end latency of around 50 ms. One such example is pre-accident sensing and generating warning messages. For lane change sensing and warning generation, end-to-end latency is specified as 100 ms.

Therefore, to fulfil such precise delay requirements, accurate timing is highly desirable among all nodes of the CR-IoV network.

Other than latency, secure communication to facilitate application-oriented tasks is also a key requirement. CR-IoV applications must be capable of opposing threats like session hijacking and jamming, which can cause serious accidents or even deaths. For this reason, overall activity needs to be monitored. The effectiveness of monitoring is highly dependent on synchronization of node.

5.6 ADVANCES TOWARDS CR-IoV TIME SYNCHRONIZATION

The time synchronization protocols discussed in Section 5.1 may not be applicable in some scenarios where nodes of the CR-IoV networks are highly mobile. Creating trees or a hierarchy might not be a good idea when neighbors keep changing very drastically. For such situations, time synchronization in the IoV is governed by IEEE 802.11p, which is an amendment of IEEE 802.11. Therefore, synchronization techniques from the IEEE 802.11 family are naturally applicable in the CR-IoV. It is always a good idea to decide which wireless time synchronization methods are applicable to the specific application under consideration and mobility of nodes involved. It could be from protocols like TPSN, FTSP, or RBS described previously or from the IEEE 802.11 family or using a GNSS approach.

5.6.1 Timing Synchronization Function–Based Synchronization in the CR-IoV

In the IEEE 802.11 standard family, a station is like a vehicular node in the CR-IoV and is attached to an access point (AP), which can be like any fixed infrastructure node. Its time synchronization methods relay on a timing synchronization function (TSF) timer, which is like a clock of the node. TSF timer is a 64-bit counter that operates on the oscillations of a quartz crystal. Time synchronization targets correction of this timer. The approach of synchronization in infrastructure-based mode is that AP transmits beacons with TSF timer values to the stations or vehicular nodes. The receiving stations or nodes adjust their TSF timers. However, in ad-hoc mode, all stations or vehicular nodes wait for a random period of time. If any beacon is received before the random time, the station or node stops waiting and transmits its own beacon with its TSF timer value as timestamp. When a station receives a beacon, it updates its own TSF timer only if the value is larger than its own timer. In this way, nodes are synchronized with the fastest node in the network. This method based on TSF provides good results but only for single-hop networks, as this approach does not address scalability and congestion issues. The node with the fastest clock is unable to send beacons to other all nodes if there is a significant increase in the number of nodes. This causes de-synchronization, known as fastest node asynchronism.

Improved techniques are required for multi-hop network time synchronization. These techniques must address the fastest node asynchronism and time partitioning problem. Time partitioning occurs if multiple clusters have internal consistency but are completely out of sync with each other. The TSF method is further exploited, and its improved forms are adaptive TSF (ATSF) and multi-hop TSF (MTSF).

5.6.1.1 Adaptive Timing Synchronization Function–Based Synchronization

The adaptive timing synchronization function is a modified method of TSF-based synchronization. In this method, the transmission frequency of the beacon is adaptive and can be adjusted to give a priority scheme to mitigate the fastest node asynchronism problem. Whenever a node receives a beacon, it compares the received timestamp with its own clock, and if the received timestamp has a greater value, it reduces its own beacon transmission frequency. Every node keeps adjusting its beacon transmission frequency. In this way, the node with fastest clock or largest value is given priority by having a higher probability of transmitting beacons.

5.6.1.2 Multi-Hop Timing Synchronization Function–Based Synchronization

The multi-hop timing synchronization function has a path-based approach. Each node must maintain a path to the node with the fastest clock. When a beacon is transmitted from the node with the fastest clock, it must reach all the other nodes, following a maintained path without being lost. The MTSF approach is a two-step process. First, all neighboring nodes create groups and find the node with the fastest clock within each group to synchronize. The node with the fastest clock from every group is considered the root node of the group. Second, root nodes are synchronized with each other to get synchronization over all the network. In this way, MTSP mitigates the fastest node asynchronism and time partitioning problem as well because it is applicable to multi-hop networks.

All of the TSF-based synchronization methods only support internal time sharing and do not synchronize with global time or time of day. Timing advertisement (TA) and timing measurement (TM) mechanisms can be used to synchronize with global time. In timing advertisement, an external clock is attached to access points to provide a global clock reference. In timing measurement, the physical layer time synchronization method is used to synchronize the AP and station and hence reduce errors.

5.6.2 Global Navigation Satellite System–Based Synchronization in the CR-IoV

The global navigation satellite system is a well-reputed utility for navigation, positioning, and timing. This technique is used as a precise time source in many wireless distributed networks. In the CR-IoV, it is already in use to determine vehicle position and velocity. However, this requires installation of devices in all nodes.

GNSS is a generic term that can be used to refer many satellite systems that offer position and time service. USA has the global positioning system, which is the most popular system. Russia has GLONASS, Europe has GALILEO, and China has its own BDS. All GNSS systems estimate position, velocity, and time using same estimation technique. Each of the available systems has a fixed time offset between them, as they all are using different reference times. However, this offset can be determined and removed. All of the systems offer worldwide and continuous service and provide global standard UTC time. Because atomic clocks are used to generate time information, the time provided by GNSS is very precise and accurate. For these reasons, GNSS is one of the most efficient systems available as a time source.

CR-IoV applications that have very strict time requirements can now be accommodated using GNSS-based time synchronization. In modern vehicles, a GNSS device is already installed for positioning and can be used for timing information. This eliminates the need for synchronization protocols that are based on message exchanges, and their performance is heavily dependent on non-deterministic transmission delays. As installing a GNSS device in every other node like street lamps can be expensive, another approach is to use the GNSS devices that are already available in vehicles and use synchronization protocols for the rest of the nodes. This hybrid model can provide cheaper time synchronization throughout the network.

Time-sharing protocols are based on message exchanges and have uncertainties, whereas GNSS time transfer has measurable and deterministic delays which can be computed straightforwardly and eliminated. GNSS path delays are small because of line-of-sight (LoS) communication between the satellite and CR-IoV node. CR-IoV nodes like vehicles and infrastructure are mostly outdoor based, and achieving LoS with a satellite is not a problem, other than in a few special cases like vehicles passing through a tunnel. Second, GNSS is less impacted by weather and noise in comparison to ground-based wireless systems. These reasons provide enough motivation for GNSS time synchronization to be used in the CR-IoV. Advanced deployment of such time synchronization systems is also available for the CR-IoV, which includes the use of multi-GNSS constellations for more precision, such as precise point positioning (PPP), differential GNSS (DGNSS), and space-based argumentation systems (SBASs).

5.6.2.1 Synchronization Using GPS Time

GPS time is a precise standard time. A precise atomic clock is used to continuously generate a timing signal. GPS time is related to UTC standard time and is ahead of UTC by 18 s. To extract time information from a satellite, a GPS receiver is required, which is an electronic device. Lots of such GPS devices are manufactured to facilitate this time transfer. Every GPS receiver device maintains timing information within itself using its own clock, which is usually a quartz clock. These GPS device clocks keep synchronizing with the satellite using GPS signals. The quartz clock update rate is 5 to 10 Hz for low-end receivers and up to 50 Hz for high-end receivers.

Among many available GPS time transfer methods, time dissemination is the simplest method. It synchronizes the time on the available GPS signal from one or more GPS satellites with an accuracy of around 40 ns. For more accurate time, the common view method is popular one. It can provide an accuracy of around 10 ns by avoiding local ionosphere and troposphere delays. This method is similar to differential GPS and is capable of determining 3D coordinates and clock offsets between two GPS receivers. If more accuracy is desirable, then the carrier-phase method can provide up to sub-nanosecond–level accuracy. However, this comes with an additional cost of dual frequency phase receivers, making it unsuitable for vehicular nodes. It uses two carrier phase signals to calculate time and achieve such high accuracy.

5.6.2.2 Mathematical Model for GNSS Time Synchronization

Devices using a GPS receiver keep themselves updated with satellite time. When a device receive a clock bias dt, then it can determine the universal time t_U by updating its time t_r. It can be mathematically written as,

$$t_U = t_r - dt - dt_U \tag{6}$$

where dt_U is the offset between universal and GPS time. Similarly, in order to synchronize two GPS receivers X and Y, receiver Y gets the clock bias with respect to receiver X, which is dt_{YX}. Now receiver Y can find its universal time as,

$$t_{YU} = t_{XU} - dt_{YX} \tag{7}$$

Now the GNSS receiver is usually equipped with the internal quartz crystal oscillator–based clock that is supposed be aligned with GPS time. Any two devices that have clocks and want to synchronize can be written as,

$$C_X(t) = d_X t + \varphi_X \tag{8}$$

$$C_Y(t) = d_Y t + \varphi_Y \tag{9}$$

Now in order to relate these two clocks, α_{XY} is the relative drift, and φ_{XY} is the phase offset. This relation can be written as,

$$C_X(t) = \alpha_{XY} C_Y(t) + \varphi_{XY} \tag{10}$$

The accuracy of synchronization between these two clocks will be increased as α_{XY} approaches 1 and φ_{XY} approaches 0.

5.6.2.3 Challenges and Solutions

Vehicular nodes in CR-IoV networks connected to GNSS may suffer from loss of navigation while driving through high-rise roads. However, it does not mean that the time signal is also lost because GPS devices can read time even from a single satellite in view, but this affects the accuracy. Some measures can be done to keep the clock synchronized in cases of total GPS signal blockage, like a vehicular node may experience while passing through a tunnel. A GPS disciplined oscillator (GPSDO) can be used, which enables the GPS device's internal oscillator to create a timing signal by mimicking past timing information. Hence GPSDO is capable of accruing a GPS signal and continuing to run the local clock of the node at a stable frequency in the absence of a GPS signal. Adaptive temperature and aging compensation can be implemented by using additional circuitry to further enhance the performance of GPSDO. However, the performance is more or less the same with or without additional compensations and can provide an accuracy of around 3 μs. This level of accuracy is sufficient for most CR-IoV applications. Another approach to synchronizing nodes that are experiencing GPS signal blockage is to synchronize them using any non–GPS-based protocols with the vehicular nodes that have active GPS connections. The nodes with active GPS connections can act as root nodes.

5.6.3 Single Timestamp–Based Cross-Layer Synchronization in the CR-IoV

In single timestamp–based cross-layer synchronization (ST-CLS) for the CR-IoV, symbol timing recovery has to be done for each node in order to communicate and understand. This happens at the physical layer, which eventually aligns the physical layer clocks of communicating nodes. Skew offset is extracted from the physical clock and is used to compensate for the skew offset of the clocks at the application layer. This requires just one timestamp in order to synchronize both the phase and frequency of the clocks. Section 5.7 gives a detailed discussion on the model and its implementation in the CR-IoV.

5.7 IMPLEMENTING TIME SYNCHRONIZATION IN THE CR-IoV

This section models synchronization in the CR-IoV using the advanced ST-CLS approach discussed previously. The main idea of this model is that the physical layer clock and application layer clock are both derived from a hardware clock (quartz based), and hence the skew estimated at the physical layer is actually the skew of the hardware clock [38]. The skew is therefore used to correct the application layer clock offset. Figure 5.16 shows the cross-layer model.

For every receiving node of the CR-IoV, the above model needs to be implemented. After the match filter, the received data undergoes interpolation using a piecewise parabolic interpolator assisted by a Modulo 1 interpolation controller. This process generates new samples at adjusted instances. A zero-crossing timing error detector (ZC-TED) and loop filter track the timing error from modified samples. Tracking and acquisition time are dependent on the loop filter parameters K_1 and K_2. Interpolation control locates the offset by updating the fractional interval. Physical layer clock skew can be extracted from the fractional interval using any estimation technique, such as least square.

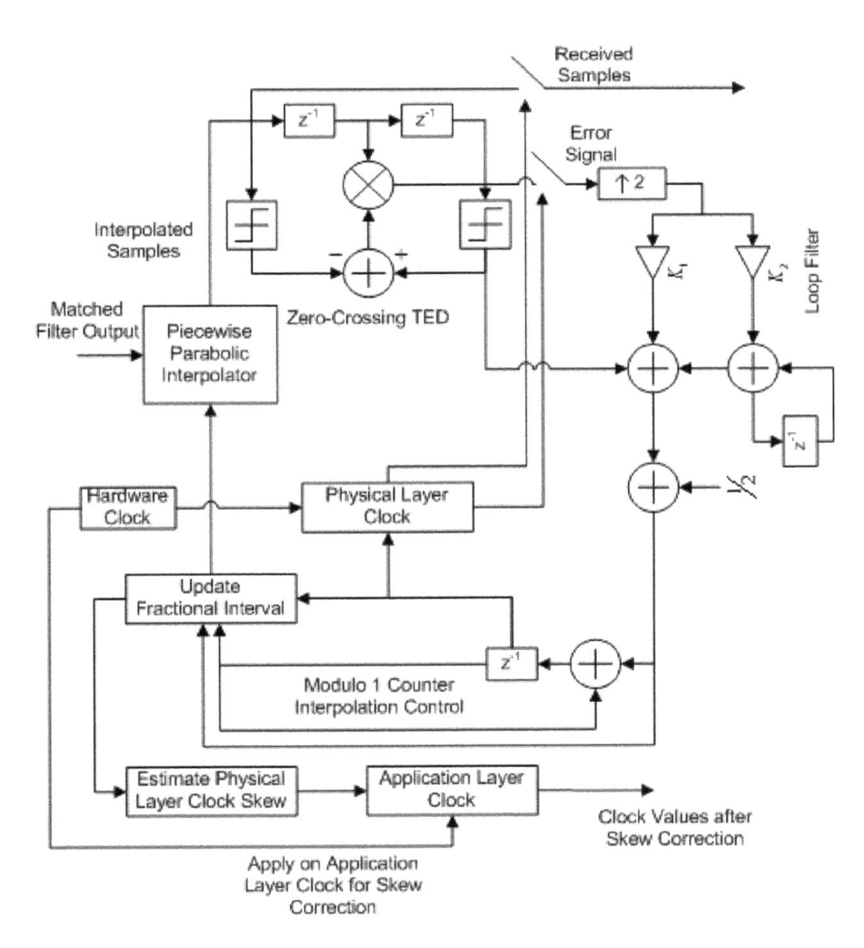

Figure 5.16 Modelling time synchronization in the CR-IoV.

5.7.1 Mathematical Model

The output of the matched filter $x(nT)$, fed to the piecewise parabolic interpolator for the k-th interpolant with basepoint index $m(k)$ and fractional change $\mu(k)$, is defined as,

$$\begin{aligned} x\Big(\big(m(k)+\mu(k)\big)T\Big) \\ &= \left(\frac{\mu(k)^3}{6}-\frac{\mu(k)}{6}\right)x\Big(\big(m(k)+2\big)T\Big) \\ &-\left(\frac{\mu(k)^3}{2}-\frac{\mu(k)^2}{2}-\mu(k)\right)x\Big(\big(m(k)+1\big)T\Big) \\ &+\left(\frac{\mu(k)^3}{2}-\mu(k)^2-\frac{\mu(k)}{2}+1\right)x\big(m(k)T\big) \\ &-\left(\frac{\mu(k)^3}{6}-\frac{\mu(k)^2}{2}+\frac{\mu(k)}{3}\right)x\Big(\big(m(k)-1\big)T\Big) \end{aligned} \tag{11}$$

The output of the interpolator processed by ZC-TED to generate the timing error signal is given by,

$$e(k) = x\Big(\big(k+\tfrac{1}{2}\big)T_s+\hat{\tau}\Big)\big[a(k-1)-a(k)\big] \tag{12}$$

where $a(k)$ and $a(k\text{-}1)$ are symbol decisions for binary PAM. The fractional interval is computed using the modulo-1 counter values η and the following equation,

$$\eta(n+1) = \big(\eta(n)-W(n)\big)\ mod\ 1 \tag{13}$$

where $W(n) = \frac{1}{N}+v(n)$, $v(n)$ is the output of the loop filter, $N = T_s/T$, T_s is the symbol time and T is the sample time. To compute the fractional interval $\mu(k)$ for the computed basepoint index $m(k)$, we have,

$$\mu\big(m(k)\big) = \frac{\eta\big(m(k)\big)}{W\big(m(k)\big)} \tag{14}$$

$\mu(k)$ and $m(k)$ are now used to compute the next interpolant. Now LS can be applied on the fractional interval to compute skew and the new sampling rate as,

$$f_s = (2 + m) f_d \tag{15}$$

where f_d is the symbol rate and m is the slope of the fractional change and physical layer clock skew. This physical layer estimate of clock skew is now applied to the application layer clock for correction, which eliminates the need of multiple timestamp exchange. Hence one timestamp is enough to synchronize CR-IoV nodes in time [39–40].

5.7.2 Simulation

MATLAB is used for simulation of two synchronization systems. The physical layer system is the implementation of the one shown in Figure 5.16. Other configurations are as follows. A binary PAM system with symbol rate 1000 symbols/sec, 8 samples per symbol, and pulse shaping using SRRC with 50% excess bandwidth [41–42]. The symbol timing recovery system estimates the physical layer clock skew at the receiver side. A skew of -1.2484×10^{-3} is extracted, which is shown in Figure 5.17.

To simulate the application layer synchronization system, the transmitter application layer clock counts the oscillation to 100,000, and the receiver application layer clock counts to 100,124 for the same duration, 500 seconds, while exchanging their clocks. The skew estimated using LS is $1 - ((100000 - 1) = (100124 - 1)) = -1.2385 \times 10^{-3}$. The result shows that the physical layer clock skew (-1.2484×10^{-3}) and application layer

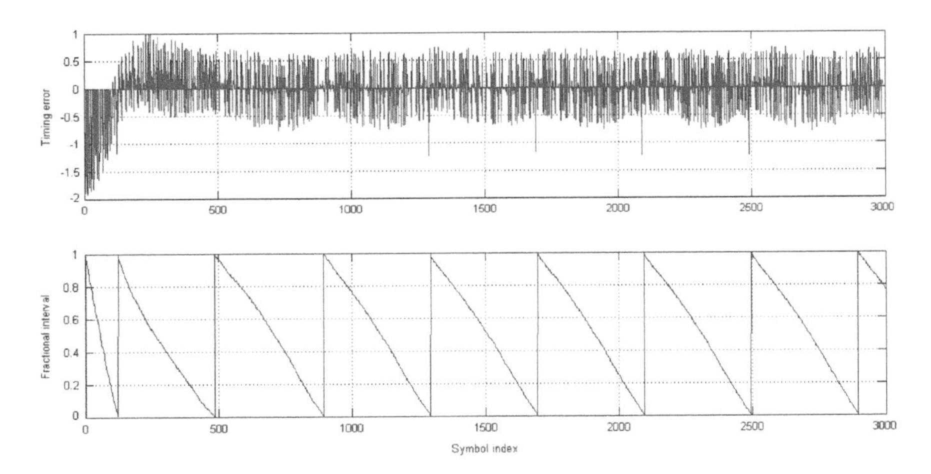

Figure 5.17 Timing error and fractional interval show skew of -1.2484×10^{-3}.

clock skew (-1.2385×10^{-3}) are approximately identical. Now if we apply a cross-layer approach, then the physical layer skew estimate can correct the application layer clock skew using one timestamp.

5.7.3 Experimentation

The experimental setup for the model uses two digital signal processing kits (TMS320C6713 DSK), DSK Tx and DSK Rx. DSK Tx is a transmitter, and DSK Rx is a receiver. These DSKs model two CR-IoV nodes that need synchronization. DSK TX generates 22,000 binary PAM symbols with 4000 symbols/sec. The sampling rate is 16,000 samples/sec, with 4 samples/symbol. Symbols are pulse shaped using SRRC and transmitted to DSK Rx. Here, DSK Rx implements the model of Figure 5.16. DSK Rx uses the same configuration along with loop filter parameters $B_n T_s = 0.005, \xi = \frac{1}{\sqrt{2}}, K_p = 2.7, K_0 = -1, N = 2$. A fractional interval is obtained and shown in Figure 5.18, and its slope reflects the skew between DSK Tx and DSK Rx.

The LS estimate is used to find the slope of the fractional interval, which is skew at the physical layer between DSK Tx and DSK Rx, as shown in Figure 5.19. The physical layer skew is estimated to be 1.3139 ppm.

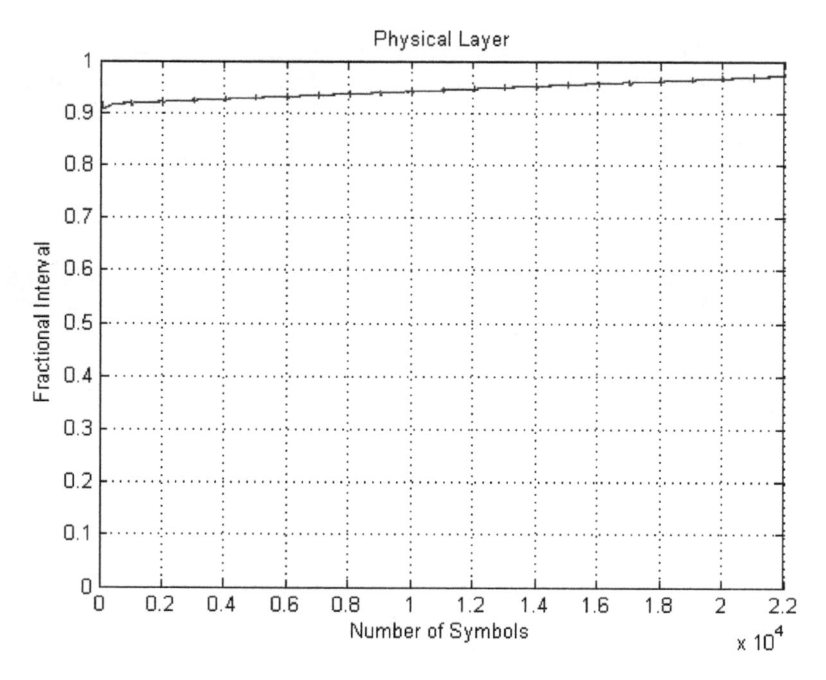

Figure 5.18 Fractional interval at the physical layer of DSK Rx.

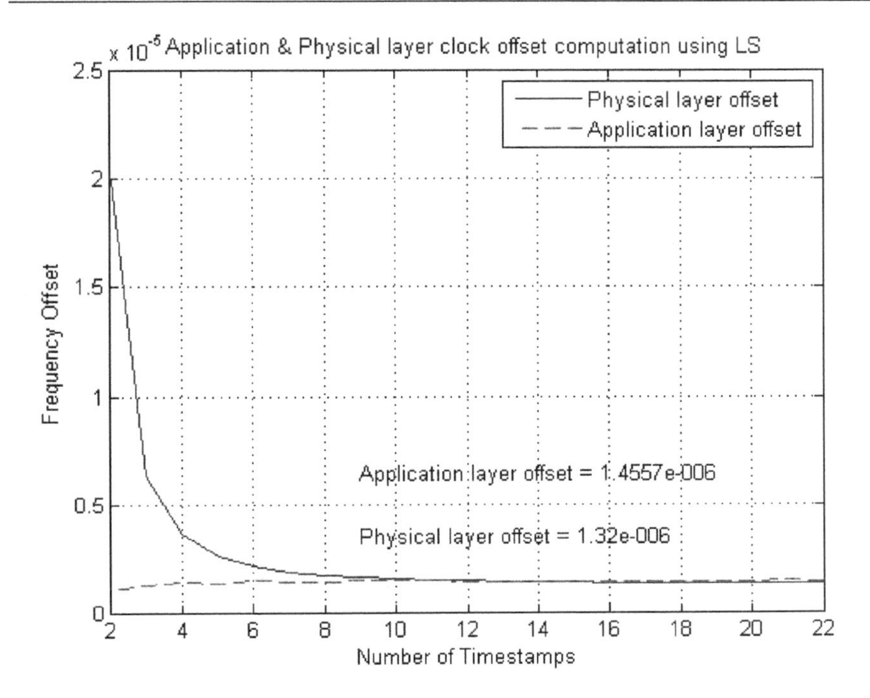

Figure 5.19 Physical and application layer clock skew.

Now to compute the application layer clock skew, the transmissions made by DSK Tx are 22 timestamps to DSK Rx. These timestamps are used to compute the skew at the application layer using LS. The estimated skew comes out be 1.4557 ppm, which is nearly the same as the physical layer offset, 1.32 ppm, as shown in Figure 5.19.

The skew at the application layer and at physical layers converges to nearly the same value. This means that we can just use one timestamp and extract the skew offset of the physical layer clock to correct the application layer clock. Multiple timestamp exchanges may not be needed.

5.7.4 Mathematical Analysis

5.7.4.1 Energy Efficiency

The previous method makes the system energy efficient, as one timestamp replaces multiple timestamp exchanges. Energy efficiency can be written as,

$$E_{eff} = \frac{N_P N_b}{N_{sym} N_{bsym}} \tag{16}$$

where N_P is the number of timestamps required to achieve skew correction by any application layer time synchronization protocol, N_b is bits in one packet, N_{sym} is total symbols, and N_{bsym} is the number of bits in one symbol.

5.7.4.2 Symbols for Cross-Layer Packet

Now if we design a such a cross-layer packet for the model of Figure 5.16 to work, then that packet must contain a specific number of symbols, defined as,

$$T_{SYM} = \frac{T_T}{\pi}\sqrt{\frac{N(N^2-1)}{\frac{2}{3}\xi\frac{E_s}{N_0}A}} + P_{SYM} \quad (17)$$

where P_{SYM} is the symbols used for timing phase synchronization, meaning it contains the timestamp value; T_T is total transmission time; N is the number of timestamps $t[n]$ if used to have the same accuracy; ξ is the loop parameter; $\frac{E_s}{N_0}$ is symbol to noise energy; and A is defined as,

$$A = \frac{N(N^2-1)}{12}\sum_{n=0}^{N-1}\left(t[n]-\bar{t}\right)^2 - \left(\sum_{n=0}^{N-1} nt[n] - \frac{N}{2}(N-1)\bar{t}\right)^2 \quad (18)$$

5.7.4.3 Clock Jitter

The clock jitter of the clocks used in the simulation and experimentation results is Gaussian, and that's why the probability density function (PDF) for the physical layer skew estimate is also Gaussian and is written as,

$$p(P) = \frac{1}{\sqrt{2\pi\sigma_P^2}} e^{-1/2\sigma_P^2 (P-\mu_P)^2} \quad (19)$$

5.8 SUMMARY

Synchronized communication in the CR-IoV is the basis for many applications. Precision and accuracy are critical in such a highly dynamic environment. Time synchronization enables coordination and consistency of events, providing real-time control and sequencing. This chapter examined the importance of time synchronization in CR-IoV networks. The existing synchronization protocols of distributed systems and their applicability were discussed for the CR-IoV; however, they may not provide excellent results for highly time-sensitive applications. GNSS-based time synchronization seems to be a promising approach if it is used in a hybrid way with the synchronization protocol of a distributed wireless network in the short-term absence of a GNSS signal. An advance cross-layer approach for CR-IoV time synchronization was modeled and implemented to show synchronizations as various layers of networks. A future research dimension could be the development of techniques that make GNSS-based time synchronization and cross-layer time synchronization more practical for CR-IoV networks.

REFERENCES

1. Alalewi, A., Dayoub, I., & Cherkaoui, S., "On 5G-V2X use cases and enabling technologies: A comprehensive survey," *IEEE Access*, 9, 2021.
2. Ali, Z., Lagen, S., Giupponi, L., & Rouil, R.. "3gpp nr v2x mode 2: Overview, models and system-level evaluation," *IEEE Access*, 9, 2021.
3. Amjad, Z., Sikora, A., Hilt, B., & Lauffenburger, J., "Low latency V2X applications and network requirements: Performance evaluation," *2018 IEEE Intelligent Vehicles Symposium (IV)*, 2018.
4. Do, D., Van Nguyen, M., Voznak, M., Kwasinski, A., & De Souza, J. N., "Performance analysis of clustering car-following V2X system with wireless power transfer and massive connections," *IEEE Internet of Things Journal*, 2021.
5. Hasan, M., Mohan, S., Shimizu, T., & Lu, H., "Securing vehicle-to-Everything (V2X) communication platforms," *IEEE Transactions on Intelligent Vehicles*, 5(4), pp. 693–713, 2020.
6. Jiang, L., Molnár, T. G., & Orosz, G., "On the deployment of V2X roadside units for traffic prediction," *Transportation Research Part C: Emerging Technologies*, 129, 103238, 2021.
7. Krajzewicz, D., Bieker, L., Härri, J., & Blokpoel, R., "Simulation of V2X applications with the iTETRIS system," *Procedia – Social and Behavioral Sciences*, 48, pp. 1482–1492, 2012.
8. MacHardy, Z., Khan, A., Obana, K., & Iwashina, S., "V2X access technologies: Regulation, research, and remaining challenges," *IEEE Communications Surveys & Tutorials*, 20(3), pp. 1858–1877, 2018.
9. Shaer, I., Haque, A., & Shami, A., "Multi-component V2X applications placement in edge computing environment," *IEEE International Conference on Communications (ICC)*, 2020.
10. Skiribou, C., & Elbahhar, F., "V2X wireless technology identification using time–frequency analysis and random forest classifier," *Sensors*, 21(13), 4286, 2021.
11. Thanh Le, T. T., & Moh, S., "Comprehensive survey of radio resource allocation schemes for 5G V2X communications," *IEEE Access*, 9, 2021.
12. Wang, J., Shao, Y., Ge, Y., & Yu, R., "A survey of vehicle to everything (V2X) testing," *Sensors*, 19(2), 334, 2019.
13. Abbasi, M., Shahraki, A., Barzegar, H. R., & Pahl, C., "Synchronization techniques in Device to device- and vehicle to vehicle-enabled cellular networks: A survey," *Computers & Electrical Engineering*, 90, 2021.

14. Gyawali, S., Xu, S., Qian, Y., & Hu, R. Q., "Challenges and solutions for cellular based V2X communications," *IEEE Communications Surveys & Tutorials*, 23(1), pp. 222–255, 2021.
15. Hasan, K. F., Feng, Y., & Tian, Y., "GNSS time synchronization in vehicular ad-hoc networks: Benefits and feasibility," *IEEE Transactions on Intelligent Transportation Systems*, 19(12), pp. 3915–3924, 2018.
16. Hasan, K. F., Wang, C., Feng, Y., & Tian, Y., "Time synchronization in vehicular ad-hoc networks: A survey on theory and practice," *Vehicular Communications*, 14, 39–51, 2018.
17. Manolakis, K., & Xu, W., "Time synchronization for multi-link D2D/V2X communication," *84th Vehicular Technology Conference (VTC-Fall)*, 2016.
18. Yoshioka, S., & Nagata, S., "Cellular V2X standardization in 4G and 5G," *IEICE Transactions on Fundamentals of Electronics, Communications and Computer Sciences*, 2021.
19. Abboud, K., Omar, H. A., & Zhuang, W., "Interworking of DSRC and cellular network technologies for V2X communications: A survey.," *IEEE Transactions on Vehicular Technology*, 65(12), pp. 9457–9470, 2016.
20. Baoguo Yang, Letaief, K., Cheng, R., & Zhigang Cao., "Timing recovery for OFDM transmission," *IEEE Journal on Selected Areas in Communications*, 18(11), pp. 2278–2291, 2000.
21. Bar-Ness, E., & Panayirci, E., "A new approach for evaluating the performance of a symbol timing recovery system employing a general type of nonlinearity," *Discovering a New World of Communications*, 1992.
22. Chen, S., Hu, J., Shi, Y., & Zhao, L., "LTE-V: A TD-LTE-Based V2X solution for future vehicular network," *IEEE Internet of Things Journal*, 3(6), pp. 997–1005, 2016.
23. D'Andrea, A., & Luise, M., "Optimization of symbol timing recovery for QAM data demodulators," *IEEE Transactions on Communications*, 44(3), pp. 399–406, 1996.
24. Dong Kyu Kim, Sang Hyun Do, Hong Bae Cho, Hyung Jin Chol, & Ki Bum Kim, "A new joint algorithm of symbol timing recovery and sampling clock adjustment for OFDM systems," *IEEE Transactions on Consumer Electronics*, 44(3), pp. 1142–1149, 1998.
25. Donghoon Lee, & Kyungwhoon Cheun., "A new symbol timing recovery algorithm for OFDM systems," *International Conference on Consumer Electronics*, 1997.
26. Gini, F., & Giannakis, G., "Frequency offset and symbol timing recovery in flat-fading channels: A cyclostationary approach.," *IEEE Transactions on Communications*, 46(3), pp. 400–411, 1998.
27. Hobert, L., Festag, A., Llatser, I., Altomare, L., Visintainer, F., & Kovacs, A., "Enhancements of V2X communication in support of cooperative autonomous driving," *IEEE Communications Magazine*, 53(12), 64–70, 2015.
28. Lim, D., "A modified Gardner detector for symbol timing recovery of M-PSK signals," *IEEE Transactions on Communications*, 52(10), pp. 1643–1647, 2004.
29. Hashmi, M. U., Hussain, M., Babar, M., & Qureshi, B., "Single-Timestamp Skew Correction (STSC) in V2X Networks," *Electronics*, 12(6), 1276, 2023.
30. Liu, G.,"A combined code acquisition and symbol timing recovery method for TDS-OFDM". *IEEE Transactions on Broadcasting*, 52(4), pp. 585–585, 2006.

31. Molina-Masegosa, R., & Gozalvez, J., "LTE-V for Sidelink 5G V2X vehicular communications: A new 5G technology for short-range vehicle-to-Everything communications," *IEEE Vehicular Technology Magazine*, 12(4), pp. 30–39, 2017.
32. Römer, K., "Time synchronization in ad hoc networks," *Proceedings of the 2nd ACM international symposium on Mobile ad hoc networking & computing – MobiHoc '01*, 2001.
33. Bregni, S., "Characterization and modelling of clocks", *Synchronization of Digital Telecommunications Networks*, pp. 203–281, 2002.
34. Hashmi, S. U., Hussain, M., Muslim, F. B., Inayat, K., & Hwang, S. O., "Implementation of symbol timing recovery for estimation of clock skew," *International Journal of Internet Technology and Secured Transactions*, 11(3), 241, 2021.
35. Elson, J., Girod, L., & Estrin, D., "Fine-grained network time synchronization using reference broadcasts," *Proceedings of the 5th symposium on Operating systems design and implementation – OSDI '02*, 2002.
36. Huang, D., Teng, W., & Yang, K., "Secured flooding time synchronization protocol with moderator," *International Journal of Communication Systems*, 26(9), pp. 1092–1115, 2013.
37. Sattar, D., Sheltami, T. R., Mahmoud, A. S., & Shakshuki, E. M., "A comparative analysis of flooding time synchronization protocol and recursive time synchronization protocol," *Proceedings of International Conference on Advances in Mobile Computing & Multimedia – MoMM '13*, 2013.
38. Rice, M. (2009). Digital communications: A discrete-time approach. Prentice Hall, 2009.
39. Yang, H., & Snelgrove, M. (n.d.). "Symbol timing recovery using oversampling techniques," *Proceedings of ICC/SUPERCOMM '96 – International Conference on Communications*, 1996.
40. U. Hashmi, M. Hussain, F. B. Muslim, K. Inayat, and S. O. Hwang, "Clock Frequency Offset Estimation Using Symbol Timing Recovery," *7th International Conference on Green and Human Information Technology*, 2019.
41. U. Hashmi, M. Hussain, N. Arshad, K. Inayat and S. O. Hwang, "Optimized Time Synchronization for Cognitive Radio Networks," *7th International Conference on Green and Human Information Technology*, 2019.
42. U. Hashmi, M. Hussain, N. Arshad, K. Inayat & S. O. Hwang, "Energy efficient cross layer time synchronisation in cognitive radio networks," *International Journal of Internet Technology and Secured Transactions*, 11(4), 329, 2021.

Chapter 6

Data Science Applications, Approaches, and Challenges in Cognitive Radio–Based IoV Systems

Mirza Anas Wahid and Syed Hashim Raza Bukhari

6.1 INTRODUCTION

With the revolution of industries and social economy, the vehicle industry is growing worldwide dramatically. In today's world, transportation in many countries is becoming heavily crowded as the population grows [1]. In many cases, the transport network is outdated and too expensive to modernize, and replacing it would be too difficult [2]. According to an ongoing report, the total number of motor vehicles used is slightly more than 1 billion, and it is expected to increase up to 2 billion by 2035 [3]. A vehicular ad hoc network (VANET) is a certain type of mobile ad hoc network used for communication between vehicles and roadside units [4]. A VANET transforms each cooperating car into a wireless router or mobile node, allowing vehicles to communicate with one another and forming a large network. As cars lose signal range and drop out of the network, other vehicles can join in, creating a mobile Internet by linking vehicles to one another. VANET only covers a narrow mobile network with mobility limits and a limited number of linked cars, according to our findings. Traffic congestion, big buildings, bad driving conduct, and complex road networks are all factors that make it difficult to utilize in large cities. The utilization of VANET is characterized by the temporary, unpredictable, and unstable nature of the engaged objects, resulting in limited and discontinuous service provision, which fails to meet the complete and sustainable needs of users. Traditional or widely adopted VANET implementations have been absent for decades, and the intended business interests have not materialized. Consequently, VANET adoption has decelerated. In contrast to VANET, the IoV encompasses two primary technological directions: vehicle networking and vehicle intelligence. Vehicle networking comprises VANET (also referred to as vehicle interconnection), vehicle telematics (also known as connected vehicles), and mobile internet as its three components, where the vehicle functions as a mobile terminal. Vehicle intelligence, on the other hand, entails enhancing the intelligence of the driver and the vehicle itself through the utilization of network technologies such as deep learning, cognitive computing, swarm computing, uncertainty artificial intelligence, and others. The term "Industry 4.0" was reportedly

 DOI: 10.1201/9781003284871-10

introduced in 2011 as a novel economic philosophy founded on high-tech innovation [5]. Consequently, the IoV focuses on the intelligent integration of individuals, vehicles, objects, and surroundings, forming a broader network that provides services for major cities or even entire countries. The IoV represents a multi-user, multi-vehicle, multi-thing, multi-network open and integrated network system characterized by excellent manageability, controllability, operability, and credibility. By leveraging the collaboration between computation and communication, such as human–vehicle collaborative awareness or swarm intelligence computation and cognition, the IoV can acquire, manage, and process extensive, complex, and dynamic data pertaining to humans, vehicles, objects, and environments, thereby enhancing the computability, scalability, and sustainability of complex network systems and information services. The ultimate goal of the IoV is to achieve deep integration of the human–vehicle–thing–environment, minimize societal costs, enhance transportation efficiency, elevate city service levels, and ensure user satisfaction and enjoyment of their vehicles. This definition clearly establishes VANET as a sub-system of the IoV. Furthermore, the IoV encompasses vehicle telematics, which refers to a connected vehicle that exchanges electronic data and provides information services, including location-based services, remote diagnostics, on-demand navigation, and audio-visual entertainment content. In the context of the IoV, vehicle telematics is merely a vehicle equipped with more advanced communication technologies, while the intelligent transportation system represents an application of the IoV.

In the Industry 4.0 era the vehicle industry is further evolving with the evolution of 5G, the Internet of Things (IoT) and cognitive radios; the CR-IoV concept emerges with the integration of emerging technologies such as software-defined networks (SDN), cognitive radio (CR) 5G, the IoT, and artificial intelligence (AI) [6]. As the frequency band gets overcrowded, the Internet of Things, which facilitates connectivity of network devices equipped with sensors, suffers from severe data exchange interference [7]. A unique cognitive radio–Internet of Things (CR-IoT) network emerges as a possible solution to the spectrum scarcity problem in conventional IoT networks by integrating cognitive radio capabilities into the IoT [8]. The CR-IoV further improves road safety, comfort, and efficiency. In IoV systems, all the vehicles act as multi-sensor objects with the capabilities of sensing, processing, computing, and communication in-car, nearby vehicles, and roadside infrastructure and units. All vehicle sensors have IP connectivity to the cloud servers and internet. These data are accessible to individuals at all levels of society via social networks, the Internet of Things, devices, and cloud computing. These data are massive in number and diversity, unstructured, complicated, and generated at a rapid pace [9]. Similarly to how electricity transformed the industrial process and home practices in the 19th century, a data-driven paradigm is at the heart of 20th-century processes and practices [10]. To analyze and visualize vast amounts of data, data science employs scientific methodologies such as data mining, data cleansing, exploratory data analysis, feature engineering, and the use of various machine learning algorithms [**11**].

6.2 DATA SCIENCE

The term "data science" was initially introduced in 1974 through Naur's publication titled *Concise Survey of Computer Methods* [12]. It was explained as "the science of dealing with data" [13]. The art of data science is gaining popularity across a broad variety of fields and professions. As a result, organizations or proposers from various origins and with varying objectives have presented quite varied perspectives [14]. Data science is a young generation of statistics, a consolidation of numerous interdisciplinary subjects, or a new body of knowledge. Data science can also be used to provide competencies and practices for the data profession, as well as to generate business strategies [15]. With the revolution of Industry 4.0, every industry such as cellular, automobile, healthcare, aviation, academia, and many more are generating huge amounts of data. This requires expertise and skills from various fields, including mathematics, statistics, computer science, and information technology. The data are vast and diverse, characterized by their unstructured nature, complexity, and rapid development [9]. A data-driven paradigm lies at the heart of 20th-century processes and practices, much as electricity transformed the industrial process and household practices in the 19th century [10]. To analyze and visualize vast amounts of data, data science employs scientific methodologies such as data mining, data cleansing, exploratory data analysis, feature engineering, and various machine learning algorithms [11]. There are multiple roles for data science professionals, such as data analysts, who are responsible for analyzing, summarizing, and visualizing data and generating reports. Data scientists are professionals responsible for cleaning and manipulating raw data to generate meaningful insights for businesses and industries. On the other hand, data engineers are software engineers who specialize in designing, building, and integrating data from diverse sources, which are then analyzed by data scientists.

6.3 DATA SCIENCE APPROACHES AND APPLICATIONS IN THE CR-IoV

The CR-IoV generates a huge amount of data with the advancement of the vehicle industry. The implementation of data science applications in the CR-IoV increases the efficiency of the system by solving many critical IoV-related problems through data science [16]. Data science plays a crucial role in various domains such as advanced analytics, data analytics, data analysis, descriptive analytics, predictive analytics, prescriptive analytics, explicit analytics, and implicit analytics, as shown in Figure 6.1. With the implication of data science in the CR-IoV, the efficiency of transportation will increase drastically [17]. With the involvement of data science in the CR-IoV, several concepts of transportation can be implemented, such as continuous driver monitoring, accident avoidance, behavior analysis, rush driving, vehicle fuel burning, theft alarming, and many more. In this section, we will discuss the data science approaches and applications in the CR-IoV [18].

6.3.1 Data Science Approaches

Professionals working in data science, who constantly seek solutions to a wide range of problems, often utilize data science methodology. The data science approach denotes the procedure for solving a certain problem. The data science approach and methods are shown in Figure 6.2. Data science processes in the CR-IoV mainly focus on six methods, problem identification, collection of data, data preprocessing, data exploring, machine and deep learning model implementation or development, and communication results in terms of reports to higher management.

Data science approaches for CR-IoV consist of multiple phases, such as CR-IoV problem identification, collecting data, data preprocessing, exploring the data, machine or deep learning, and communicating results.

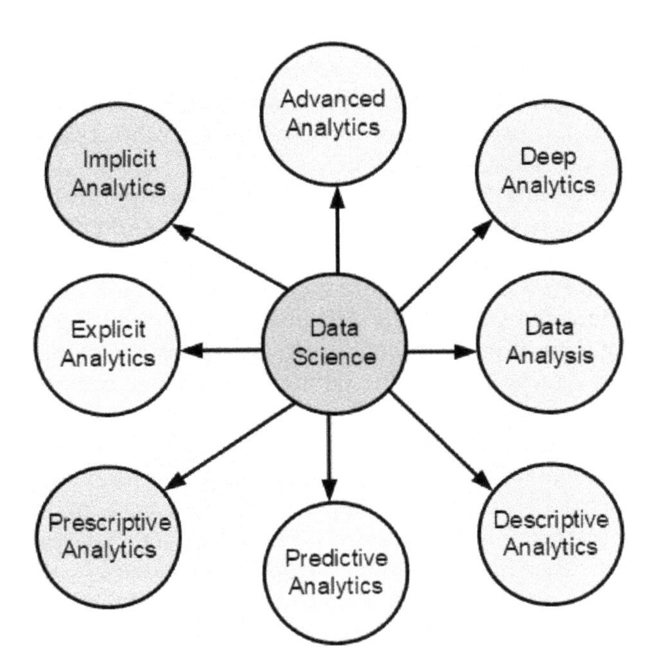

Figure 6.1 Data science empowering the CR-IoV: unleashing opportunities.

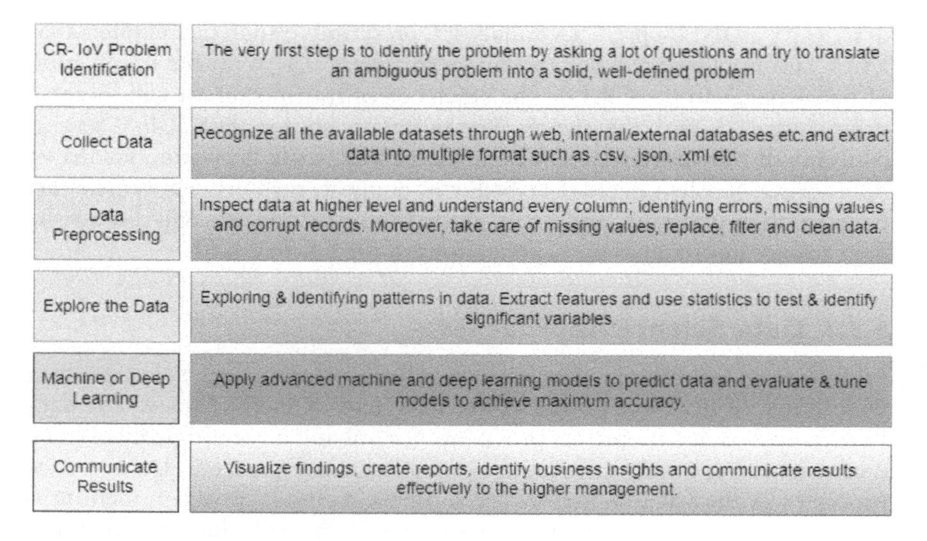

Figure 6.2 Exploring cutting-edge data science approaches in the CR-IoV.

CR-IoV Problem Identification In the first phase, the CR-IoV problem identification step is processed in which the problem is identified by asking a lot of questions and converting cryptic statements or problems into concrete, well-defined problems.

Collect Data After the CR-IoV problem identification process, the next process is the collection of data that recognizes all the available datasets from multiple platforms, such as databases and web services in different formats such as XML, .csv, and JSON.

Data Preprocessing Data preprocessing is an essential technique in data mining that aims to convert raw data into a useful and efficient format. It involves several sub-steps, including data cleaning, data transformation, and data reduction. Raw data often contain irrelevant and missing components, which are addressed through data cleansing. Data cleansing involves handling missing data, noisy data, and other related issues. Noisy data refers to irrelevant or incomprehensible data that machines cannot interpret. It can arise from inaccurate data collection or input errors. Various methods, such as regression, clustering, and binning, can be applied to address noisy data. The binning technique is employed to smooth sorted data by dividing it into equal-sized segments and applying various methods independently within each segment. To complete the task, the data in a segment can be replaced with either the mean value or boundary values. In regression, the data can be smoothed by fitting them to a regression function. This regression approach can involve a linear model with a single independent variable or a multiple regression model with multiple independent variables. Clustering methods group similar data points into clusters, potentially leaving outliers undetected or placing them outside the identified clusters. Once missing values and noisy data have been addressed in the data preprocessing stage, the subsequent step is data transformation. This step aims to convert the data into suitable forms for the mining process. Data transformation encompasses various techniques such as normalization, attribute selection, discretization, and concept hierarchy generation. Following data transformation, the subsequent process is data reduction. Since data mining deals with vast amounts of data, analysis becomes challenging. To mitigate this challenge, data reduction techniques are employed. These techniques aim to enhance storage efficiency and reduce data storage and analysis costs. The steps involved in data reduction include data cube aggregation, attribute subset selection, numerosity reduction, and dimensionality reduction.

Data Exploration Data exploration is the initial phase of data analysis, during which data analysts employ data visualization and statistical tools to characterize dataset features such as size, quantity, and accuracy. This process enhances comprehension of the data's nature. Both manual analysis and automated software solutions are employed for visual exploration, detecting relationships between data variables, dataset structure, outliers, and data value distribution. These activities unveil patterns and points of interest, enabling data analysts to gain profound insights into the raw data. Data are often gathered in vast, unstructured quantities from diverse sources, necessitating a comprehensive understanding before extracting pertinent data for subsequent analyses, including univariate, bivariate, multivariate, and principal components analysis.

Machine or Deep Learning Following data exploration, the subsequent step involves machine or deep learning. Machine learning is a specialized area of artificial intelligence that focuses on the objective of enabling computers to perform tasks without explicit programming. Typically, computers are provided with structured data and "learn" to improve their analysis and decision-making capabilities over time. Structured data refers to data organized in columns and rows, such as a food category column with row items like "fruit" or "meat" in Excel. This type of structured data are easily processed by computers, and the benefits are apparent. It is no coincidence that one of the most prominent data programming languages is called Structured Query Language (SQL).

Once programmed, a computer can continuously accept new data, categorize them, and take action without the need for additional human interaction. Even if the categorization process is stopped, the computer can eventually recognize that "fruit" falls under the category of food. This level of self-sufficiency is crucial in machine learning and distinguishes it into subcategories based on the degree of ongoing human assistance required. Supervised learning, as a sub-type of machine learning, necessitates the highest level of continual human input, as implied by its name. In supervised learning, the computer is provided with training data and a specific model designed to "teach" it how to respond to the data. This supervision helps refine the model over time, enabling it to handle new datasets that conform to the learned patterns. However, continuously monitoring and adjusting the computer's performance is inefficient. Semi-supervised learning, on the other hand, involves providing the computer with a mix of correctly labeled and unlabeled data, allowing it to autonomously search for patterns. The labeled data serve as guidance from the programmer, but no continuous adjustments are provided.

Unsupervised learning expands on this concept by utilizing unlabeled data. The machine is allowed to discover patterns and relationships on its own, often uncovering insights that a human data analyst might have overlooked. Clustering is a popular application of unsupervised learning, where the computer organizes the data into groups based on common themes and patterns it identifies. This technique is commonly employed by shopping/e-commerce websites to determine personalized recommendations for customers based on their previous purchases. In both supervised and unsupervised learning, there are no immediate consequences for the computer if it fails to analyze or categorize the data correctly. However, what if, similar to a child in school, the computer received positive feedback when it made correct predictions and negative feedback when it made errors? In such a scenario, the computer would likely start learning how to perform specific tasks through trial and error, recognizing it is on the right track when it receives a reward (e.g., a score) that reinforces its "positive behavior". This type of reinforcement learning is crucial in helping robots master complex tasks that require vast amounts of highly adaptable and unpredictable datasets. It opens up possibilities for computers to move towards achieving goals such as performing surgery or driving a car.

Machine learning enables computers to perform tasks without explicit programming, but their behavior remains machine-like. They fall short of human capabilities, particularly in challenging tasks like extracting information from images or videos. Deep learning models, inspired by the structure of the human brain, offer a highly advanced approach to machine learning that addresses these challenges. Deep neural networks consist of intricate, multi-layered structures where data are transmitted between nodes, resembling neurons, in highly interconnected ways. This allows for non-linear transformations of the data, progressively abstracting the information.

Communicate Results The last step of data science approaches in the CR-IoV is to create reports, visualize findings, identify business insights, and communicate results to higher management.

6.3.2 Applications of the CR-IoV

Recent advancements in cognitive radio technology have paved the way for the integration of CR into Internet of Vehicles systems. This integration allows for opportunistic spectrum sensing and hopping, enabling vehicles to adaptively utilize the available spectrum based on their requirements and the prevailing operating conditions. However, it is crucial to acknowledge the distinctive characteristics of vehicular communication networks when designing CR-assisted IoV systems. Factors such as high vehicle speeds, high urban area density, and dynamic topology pose challenges in selecting appropriate network configurations, communication channels, and data rates. These considerations are vital to ensure efficient and reliable communication within vehicular environments.

With the rapid advancements in technology, particularly wireless networks and the Internet of Things, the concept of the Internet of Vehicles has emerged as a highly promising paradigm. The IoV extends beyond traditional vehicular ad-hoc networks and encompasses a broader ecosystem that includes not only vehicles but also individuals, roadways, parking lots, and municipal infrastructure. It establishes real-time communication and connectivity among these various entities, enabling seamless information exchange and interaction. This interconnectedness within the IoV holds great potential for enhancing transportation systems and enabling intelligent services and applications in the context of vehicular environments [**19**]. The IoV presents an extensive and comprehensive range of applications that can be broadly categorized into safety, transportation efficiency, information/entertainment/convenience, and logistics [**19**]. Within each of these groups, there are various applications that contribute to the widespread adoption of the IoV paradigm. Figure 6.3 lists a few CR-IoV applications

CR-IoV Transport Applications In the realm of connected vehicles, safety features involve nodes transmitting real-time collision data, including position information, to emergency services. This facilitates a quicker response time, potentially saving lives. Cooperative collision warning systems can

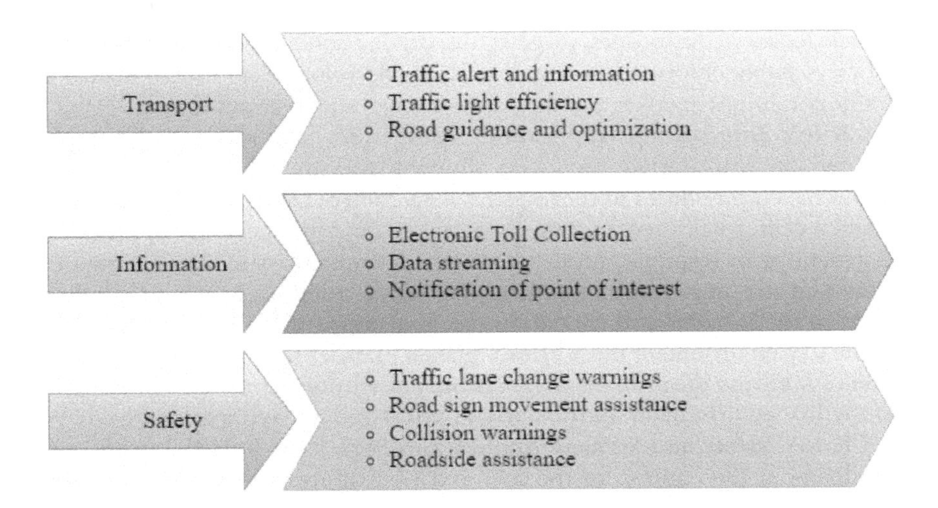

Figure 6.3 Exploring CR-IoV applications in autonomous vehicle networks.

detect potential collisions and alert the driver to take evasive action. Similarly, cooperative forward collision warning systems monitor the distance between vehicles and notify the driver to prevent rear-end collisions [20]. Driving through an intersection poses significant challenges for drivers due to the convergence of multiple traffic streams, which escalates the risk of collisions [21]. Few measures have been presented to avoid junction collisions. [22] proposes an abstraction-based approach for avoiding collisions. In [23], an algorithm based on time slots that can manage a high number of cars is presented. One of the most sought-after IoV safety applications is intelligent intersections. The Vehicle Safety Consortium (VSC) has identified various safety-related applications, including curve speed warning, pre-crash sensing, cooperative forward collision warning, emergency brake light warning, left turn assistant, traffic signal violation, lane change warning, and stop sign movement assistant [19]. The IoV can be utilized to develop transportation efficiency applications, including route guidance and optimization, as well as green light efficiency. Certain applications within this domain may require the presence of roadside infrastructure, while others can rely solely on vehicle-to-vehicle communication [19]. The implementation of IoV applications will not only enhance traffic management but also lead to substantial reductions in trip time, fuel consumption, and pollution levels by alleviating traffic congestion. The IoV is poised to play a crucial role in the realization of the smart city concept, driving its implementation forward.

The logistics sector stands to gain significant benefits from the IoV, particularly due to its reliance on road transportation for efficient and timely product delivery. The integration of the IoV in logistics can have profound implications. Smart fleet management in logistics can harness the power of the IoV to optimize operations and enhance efficiency [24]. In the field of logistics, sensors will be installed in trucks to collect a wide range of data, including tire pressure, vehicle location, diagnostics, cargo status, driving conditions, temperature, and more. A public information center will provide valuable information such as traffic conditions and weather forecasts, among other relevant details.

CR-IoV Information Applications RFIDs and barcodes will be implemented on transported products, allowing for the transmission of data about the commodities to the logistics data control center. The logistics data control center will then analyze the received data and generate appropriate instructions in response. Security and the level of trust in the information conveyed are important considerations when building transport applications in the IoV. An authentication mechanism for distributed information in the IoV environment must be established in order to prevent the propagation of fake and illegitimate information. False information may be harmful in the IoV environment, particularly in safety-related applications.

CR-IoV Safety and Security Applications Security is indeed a prominent challenge in the context of the IoV and necessitates further investigation. Within the realm of IoV, entertainment/information/convenience applications encompass various aspects such as content streaming to in-car entertainment systems and point-of-interest announcements for amenities like gas pumps, restaurants, and bathrooms. Additionally, convenience applications include electronic toll deductions at tolling stations, relay of vehicle diagnostics to manufacturers for automated service scheduling, web browsing, streaming services, and roadside assistance offerings such as location and price information for fuel stations, parking, and restaurants, among others. These applications contribute to enhancing convenience and entertainment within the IoV environment [25].

6.4 CHALLENGES AND ISSUES

In this section, we will delve into various issues and challenges related to the implementation of data science in the CR-IoV. These matters will aid in identifying future research avenues in the realm of data science in the CR-IoV. Two approaches exist to address primary research challenges: one involves delineating concerns raised by pertinent groups, while the other entails scrutinizing the intricacies and characteristics of data science quandaries as complex systems. Data has emerged as the novel lifeblood of enterprises, assuming a pivotal role in all decision-making procedures. Currently, a wide array of industries relies on data and analytics to fortify their market standing and bolster revenue streams.

6.4.1 Understanding of Learning

As the use of analytics methodologies such as data science and big data analytics has grown, so have the issues in data science. The majority of data science concerns are not company specific. These challenges may involve acquiring the proper people or resolving basic concerns such as organizing raw data, unforeseen security risks, and so on. In Figure 6.4, the data science challenges of the CR-IoV are discussed. As much as we respect deep learning's astounding results, we still lack a scientific explanation of why deep learning works so effectively, although we are making progress. We don't fully comprehend the mathematical characteristics of deep learning algorithms or the models they generate [26]. The exact reasons behind the divergent outcomes generated by a deep learning model remain elusive. The resilience or susceptibility of models to alterations in input data distributions remains uncertain. Ensuring optimal performance of deep learning models on novel input data poses a challenge. Describing or quantifying the uncertainty inherent in a model's output remains an area of limited understanding.

6.4.2 Computing Systems for Data-Intensive Applications

Deep learning's underlying computational boundaries are unknown to us [27]. When do more data and more computation no longer help? Deep

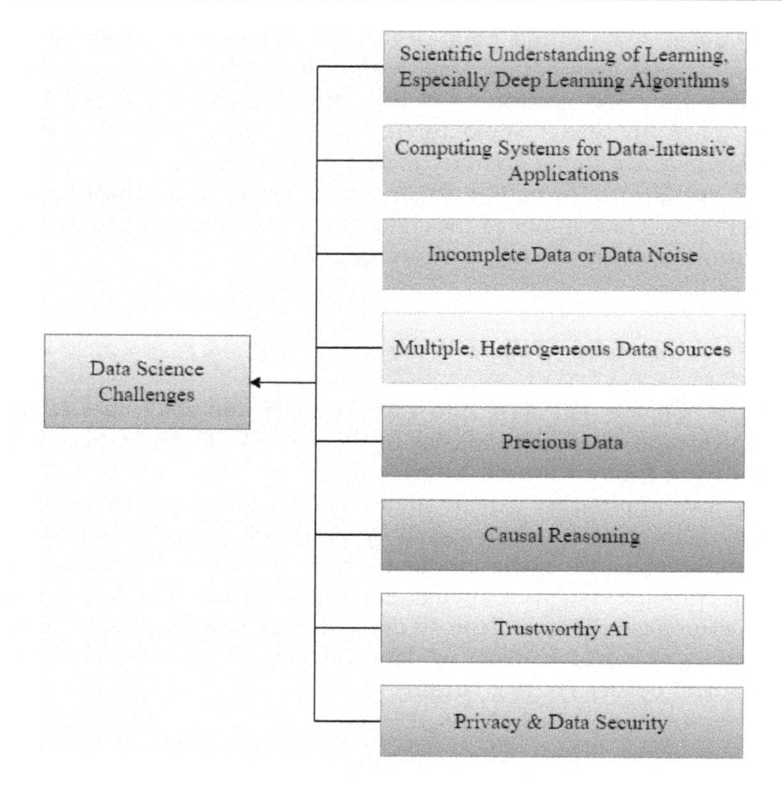

Figure 6.4 Interpreting complex CR-IoV data: challenges for data scientists.

learning is an example of a topic where experimentation takes precedence over perfect theoretical understanding. Furthermore, it is not the only example in learning: random forests [28] and high-dimensional sparse statistics are also examples [29]. Computing systems are also one of the challenges for data-intensive applications; traditional computer system designs have prioritized processing speed and power: the more cycles, the faster the program can operate. Today, data are the major focus of applications, particularly in the sciences (for example, astronomy, biology, climate science, and materials science). GPUs, FPGAs, and TPUs are examples of novel special-purpose processors that are now routinely found in major data centers. Domain-specific accelerators, particularly those developed for deep learning, outperform general-purpose processors by orders of magnitude [30]. Even with all of these data and all of this rapid and flexible processing capacity, it might take weeks to construct good predictive models; yet, applications in research and business require real-time forecasts. Distributing data, computers, and models aids in scalability and dependability (as well as privacy) but comes up against the basic constraint of light speed and practical network capacity and latency restrictions. Furthermore, data- and computation-hungry methods, such as deep learning, consume a lot of energy [31]. In our performance measures, we should account for not just area and time but also energy usage. In summary, we must rethink computer system architecture from the ground up, with data (rather than computation) as the primary focus. When designing new computer systems, it is imperative to consider various factors such as heterogeneous processing capabilities; efficient data layout for rapid access; communication and network capacity; energy efficiency; and the specific requirements of the target domain, application, or task. These considerations are crucial for optimizing system performance, resource utilization, and overall effectiveness in meeting the intended goals and objectives.

6.4.3 Incomplete Data or Data Noise

It is very challenging to extract information from incomplete data. The real world is inherently chaotic, and our understanding of every individual data point is often incomplete. However, data scientists strive to leverage such data to build models that facilitate prediction and inference. This long-standing statistical challenge arises due to several key factors. First, the volume of data, especially concerning individuals, continues to expand exponentially, presenting new complexities. Second, the process of data generation and collection is not entirely within our control, resulting in variations across different users and populations, as seen in data from mobile phone and web applications designed with specific user characteristics in mind. Last, numerous sectors, including finance, retail, and transportation, are increasingly focused on achieving real-time personalization, driving the need for efficient data analysis and tailored services. The intended use of differential privacy for Census 2020 data [32], in which noise is purposely added to a query result to safeguard the anonymity of people participating in the census, is an excellent example of a fresh articulation of this topic. Addressing "intentional" noise becomes particularly crucial for researchers working with small geographic regions like census blocks, as the presence of additional noise can render the data less informative at such fine levels of aggregation. Social scientists, who have long relied on census data to draw conclusions, face the challenge of deriving meaningful inferences from these "noisy" data and integrating their previous findings with the new ones. The advent of machine learning provides an opportunity to enhance the efficiency and accuracy of these judgments by enabling noise filtering techniques that distinguish signals from the noise in the data. By leveraging machine learning capabilities, social scientists can mitigate the impact of noise, leading to more effective and precise analysis and facilitating the integration of past and new findings.

6.4.4 Heterogeneous Data Sources

Data scientists often face challenges when dealing with diverse data sources, making it difficult to extract valuable insights. This can result in spending considerable time on data filtering, leading to potential errors and unreliable decision-making. In such scenarios, the standardization of data becomes crucial to ensure accurate analysis. To determine the appropriate format for data science (DS), it is essential to grasp the fundamentals of big data.

6.4.5 Privacy and Data Security

Data security is a critical concern in the field of data science, impacting businesses worldwide. It encompasses a wide range of security measures and tools employed in analytics and data processes. Some common data security breaches include attacks on data systems, instances of ransomware, and data theft. Information theft poses a significant risk, especially for organizations handling sensitive data like financial information or personal customer information. With the exponential growth of data exchange over networks, the threat to data during transmission has increased substantially. To mitigate these risks, businesses must prioritize three fundamental aspects of data security: confidentiality, integrity, and accessibility.

6.4.6 Precious Data

Data hold value for three main reasons: they may be costly to gather; they may involve rare occurrences with a low signal-to-noise ratio; or they may be artisanal, tailored to specific tasks or a niche audience. Expensive data can be exemplified by large, unique, and costly scientific tools. Each type of valuable data requires distinct data science methodologies and algorithms that consider the domain, intended purposes, and users of the data. Machine learning is a powerful technique for uncovering patterns, exploring relationships, and identifying correlations within massive datasets. While the application of machine learning has expanded the horizons of research in fields such as economics, social science, public health, and medicine, these disciplines require methodologies that go beyond mere correlation studies and can effectively address causal inquiries.

6.4.7 Causal Reasoning

The study of causal inference in the context of enormous volumes of data is a rich and expanding topic of contemporary research. Economists are developing new ways for incorporating the quantity of data presently accessible into their traditional causal reasoning approaches, such as the use of instrumental variables; these new methods make causal inference estimation more efficient and flexible [33]. Data scientists are increasingly delving into multiple causal inference, recognizing that real-world observations are often influenced by multiple interacting factors. This shift in focus allows researchers to move beyond the constraints of univariate causal inference and explore more complex relationships. By considering the interplay of multiple factors, data scientists aim to overcome the limitations posed by strong assumptions and gain a deeper understanding of the intricate dynamics that shape real-world phenomena. This expansion into multiple causal inference opens up new avenues for analysis and provides a more comprehensive perspective on the complexities of causal relationships [34]. With the increasing availability of government agency and commercial data to the public, data scientists are applying synthetic control methodologies for creative implementations in public health, retail, and sports. These applications draw inspiration from natural experiments conducted in the fields of economics and the social sciences [35].

6.4.8 Trustworthy AI

In various critical domains like autonomous vehicles, criminal justice, health care, hiring, housing, human resource management, law enforcement, and public safety, the utilization of artificial intelligence and machine learning systems has experienced swift advancement. These systems play a significant role in decision-making processes that directly affect human lives. To foster trust in these systems, one approach is to offer explanations for the outcomes produced by machine-trained models [36].

6.5 CONCLUSION

The IoV is quickly becoming a critical component of the smart city idea. It is a one-of-a-kind implementation of the Internet of Things concept. It is a complicated network system made up of disparate devices and networks that communicate with one another. As significant and valuable as the CR-IoV and data science are, they are not without limitations and obstacles, as previously noted. The deployability and spread of the IoV are dependent on addressing the concerns and obstacles that these highly dynamic networks encounter. Security and privacy are two critical elements that must be prioritized. The discussion concluded by highlighting potential research directions in the field of data science in the CR-IoV, including approaches, applications, issues, and challenges. To investigate key research challenges, two approaches can be taken. The first approach involves describing the concerns expressed by relevant groups, taking into account their specific perspectives and needs. The second approach involves examining the issues from the perspective of the inherent complexity and nature of data science problems as complex systems. Both approaches are valuable in gaining a comprehensive understanding of the research challenges and paving the way for advancements in the field of data science in the CR-IoV.

REFERENCES

1. G. D. Erhardt, S. Roy, D. Cooper, B. Sana, M. Chen, and J. Castiglione, "Do transportation network companies decrease or increase congestion?" *Science Advances*, vol. 5, no. 5, p. eaau2670, 2019.
2. K. Richly, J. Brauer, and R. Schlosser, "Predicting location probabilities of drivers to improve dispatch decisions of transportation network companies based on trajectory data," in *ICORES*, 2020, pp. 47–58.
3. F. Sakiz and S. Sen, "A survey of attacks and detection mechanisms on intelligent transportation systems: VANETs and IoV," *Ad Hoc Networks*, vol. 61, pp. 33–50, 2017.
4. O. S. Al-Heety, Z. Zakaria, M. Ismail, M. M. Shakir, S. Alani, and H. Alsariera, "A comprehensive survey: Benefits, services, recent works, challenges, security, and use cases for SDN-VANET," *IEEE Access*, vol. 8, pp. 91028–91047, 2020.
5. J. Prinsloo, S. Sinha, and B. von Solms, "A review of Industry 4.0 manufacturing process security risks," *Applied Sciences*, vol. 9, no. 23, p. 5105, 2019.
6. B. Meindl, N. F. Ayala, J. Mendon, and A. G. Frank, "The four smarts of Industry 4.0: Evolution of ten years of research and future perspectives," *Technological Forecasting and Social Change*, vol. 168, p. 120784, 2021.
7. H. A. B. Salameh, S. Al-Masri, E. Benkhelifa, and J. Lloret, "Spectrum assignment in hardware-constrained cognitive radio IoT networks under varying channel-quality conditions," *IEEE Access*, vol. 7, pp. 42816–42825, 2019.
8. A. A. Khan, M. H. Rehmani, and A. Rachedi, "Cognitive-radio-based Internet of Things: Applications, architectures, spectrum related functionalities, and future research directions," *IEEE Wireless Communications*, vol. 24, no. 3, pp. 17–25, 2017.
9. C. Lokhorst, R. De Mol, and C. Kamphuis, "Invited review: Big data in precision dairy farming," *Animal*, vol. 13, no. 7, pp. 1519–1528, 2019.
10. K. Van Es and M. T. Schäfer, *The datafied society. Studying culture through data*. Amsterdam University Press, 2017.
11. S. Virkus and E. Garoufallou, "Data science from a library and information science perspective," *Data Technologies and Applications*, vol. 53, no. 4, pp. 422–441, 2019.
12. J. Belzer, "Concise survey of computer methods. Peter Naur. New York: Petrocelli books, 397 p. (1975)," *Journal of the American Society for Information Science*, vol. 27, no. 2, pp. 125–126, 1976.
13. H. Harris, S. Murphy, and M. Vaisman, *Analyzing the analyzers: An introspective survey of data scientists and their work*. O'Reilly Media, Inc., 2013.
14. L. Cao, "Data science: A comprehensive overview," *ACM Computing Surveys (CSUR)*, vol. 50, no. 3, pp. 1–42, 2017.
15. W. Van Der Aalst, "Data science in action," in *Process mining*. Springer, 2016, pp. 3–23.
16. P. Rawat, K. D. Singh, and J. M. Bonnin, "Cognitive radio for M2M and Internet of Things: A survey," *Computer Communications*, vol. 94, pp. 1–29, 2016.
17. S. Saponara, S. Giordano, and R. Mariani, "Recent trends on IoT systems for traffic monitoring and for autonomous and connected vehicles," *Sensors*, vol. 21, no. 5, p. 1648, 2021.
18. P. Siarry, A. K. Sangaiah, Y.-B. Lin, S. Mao, and M. R. Ogiela, "Guest editorial: Special section on cognitive big data science over intelligent IoT networking

systems in industrial informatics," *IEEE Transactions on Industrial Informatics*, vol. 17, no. 3, pp. 2112–2115, 2020.
19. H. Hartenstein and L. Laberteaux, "A tutorial survey on vehicular ad hoc networks," *IEEE Communications Magazine*, vol. 46, no. 6, pp. 164–171, 2008.
20. H. Kowshik, D. Caveney, and P. Kumar, "Provable systemwide safety in intelligent intersections," *IEEE Transactions on Vehicular Technology*, vol. 60, no. 3, pp. 804–818, 2011.
21. Y. Toor, P. Muhlethaler, A. Laouiti, and A. De La Fortelle, "Vehicle ad hoc networks: Applications and related technical issues," *IEEE Communications Surveys & Tutorials*, vol. 10, no. 3, pp. 74–88, 2008.
22. A. Colombo and D. Del Vecchio, "Supervisory control of differentially flat systems based on abstraction," in *2011 50th IEEE Conference on Decision and Control and European Control Conference*, IEEE, 2011, pp. 6134–6139.
23. A. Colombo and D. Del Vecchio, "Efficient algorithms for collision avoidance at intersections," in *Proceedings of the 15th ACM International Conference on Hybrid Systems: Computation and Control*, 2012, pp. 145–154.
24. N. Sharma, N. Chauhan, and N. Chand, "Smart logistics vehicle management system based on Internet of Vehicles," in *2016 Fourth International Conference on Parallel, Distributed and Grid Computing (PDGC)*, IEEE, 2016, pp. 495–499.
25. E. Hossain, G. Chow, V. C. Leung, R. D. McLeod, J. Mišić, V. W. Wong, and O. Yang, "Vehicular telematics over heterogeneous wireless networks: A survey," *Computer communications*, vol. 33, no. 7, pp. 775–793, 2010.
26. R. Balestriero *et al.*, "A spline theory of deep learning," in *International Conference on Machine Learning*, PMLR, 2018, pp. 374–383.
27. S. Bangalore *et al.*, "St-segment elevation in patients with covid-19–a case series," *New England Journal of Medicine*, vol. 382, no. 25, pp. 2478–2480, 2020.
28. E. Scornet, G. Biau, and J.-P. Vert, "Consistency of random forests," *The Annals of Statistics*, vol. 43, no. 4, pp. 1716–1741, 2015.
29. I. M. Johnstone and D. M. Titterington, "Statistical challenges of high-dimensional data," *Philosophical Transactions of the Royal Society: A Mathematical, Physical and Engineering Sciences*, vol. 367, no. 1906, pp. 4237–4253, 2009.
30. W. J. Dally, Y. Turakhia, and S. Han, "Domain-specific hardware accelerators," *Communications of the ACM*, vol. 63, no. 7, pp. 48–57, 2020.
31. E. Strubell, A. Ganesh, and A. McCallum, "Energy and policy considerations for deep learning in NLP," *arXiv preprint arXiv:1906.02243*, 2019.
32. M. B. Hawes, "Implementing differential privacy: Seven lessons from the 2020 United States Census," *Harvard Data Science Review*, vol. 2, no. 2, 2020.
33. S. Athey and G. W. Imbens, "Machine learning methods that economists should know about," *Annual Review of Economics*, vol. 11, pp. 685–725, 2019.
34. Y. Wang and D. M. Blei, "The blessings of multiple causes," *Journal of the American Statistical Association*, vol. 114, no. 528, pp. 1574–1596, 2019.
35. Y. Liu *et al.*, "EUDAQ2–a flexible data acquisition software framework for common test beams," *Journal of Instrumentation*, vol. 14, no. 10, p. P10033, 2019.
36. A. Adadi and M. Berrada, "Peeking inside the black-box: A survey on explainable artificial intelligence (XAI)," *IEEE Access*, vol. 6, pp. 52138–52160, 2018.

Section 5

Security and Quality of Service

Chapter 7

Security Threats and Counter Measures in the Internet of Vehicles

Muhammad Naveed Younis, Syed Hashim Raza Bukhari, and Mudassar Naseer

7.1 INTRODUCTION

The desire for information sharing and connectivity has pushed the world towards the Internet of Things (IoT) [1]. The IoT has opened the door for new possibilities like smart homes, smart cities, and the Internet of Vehicles (IoV) [2, 3] (to name a few). By 2025, there will likely be more than 75.44 billion IOT connections [4]. According to global predictions made by Statista, 70% of vehicles will be connected to the internet by 2023. The IoV aims to reduce accidents, help vehicles find the best routes, avoid traffic congestion, monitor road status, provide infotainment services, and maintain peer-to-peer connectivity in smart cities [5, 6]. With the quick service that we expect from the IoV, there are security threats which can affect these services, either causing a data breach, slowing services, or making them unreachable, and they can endanger valuable lives. These security threats pose a severe threat to valuable lives as well as to the IoV industry [7].

These threats can occur in any type of IoV communication. IoV communication can broadly be grouped into three categories: Vehicle-to-Vehicle (V2V), Vehicle-to-Infrastructure (V2I), and Vehicle-to-Cloud (V2C) [7]. Owing to the severity of the implications posed by security threats, there is a dire need for updated survey papers. This will help to find recent issues and challenges in the security of the IoV and to adopt counter measures. In our study, the most recent survey paper is [8], but this only discusses security threats and network architecture. [**6**, **7**, **9**] cover threats and countermeasures; however, they were published in 2017, 2018, and 2019, respectively. There is quite advancement in both attacking and exploiting IoV. [**10–13**] encompass only threats to the IoV. This chapter will be bounded by the classical network security goals: confidentiality, integrity, and availability. Figure 7.1 shows network security goals.

7.2 SECURITY THREATS IN THE IoV

The IoV enhances connectivity among vehicles, the infrastructure, and the cloud. However, the dynamic nature of communication in the IoV where

DOI: 10.1201/9781003284871-12

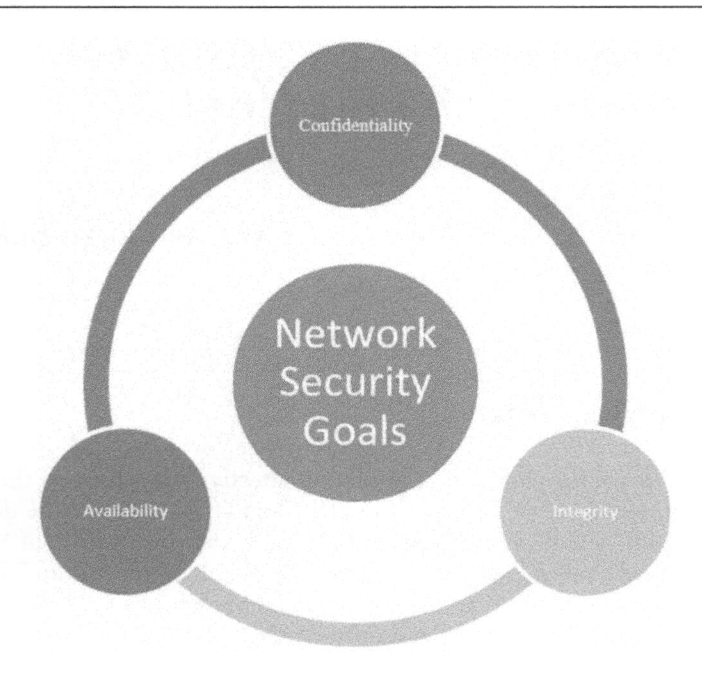

Figure 7.1 Network security goals.

data is flowing around in different aspects (vehicles, the infrastructure and the cloud) can encourage adversaries to take advantage. There are two types of attacks, active attacks and passive attacks. In active attacks, an adversary tries to modify the contents of the message being passed over the network. In passive attacks, the adversary does not modify the contents of the message, so their detection is more difficult then active attacks [9]. The adversary can generate attacks breaching confidentiality, integrity, and availability of the IoV network or user.

7.2.1 Attacks on Confidentiality of IoV Communication

Confidentiality is the property of not making information public or disclosing it to unwanted individuals, entities, or procedures. It is maintained using encryption. Table 7.1 summarizes the attacks on confidentiality in the IoV.

Table 7.1 Attacks on Confidentiality in the IoV

Motive of Attack	*Types of Attacks*	*Weakness of Attacks*
To capture and reveal the message content of communications travelling in the IoV.	• Eavesdropping • Identity exposure • Traffic analysis	Strong encryption can make it impossible for an adversary to decrypt the message content.

7.2.1.1 *Eavesdropping*

Eavesdropping (ED) is used by the adversary to listen to or capture data while the IoV communicates. Adversary can fabricate the messages or transmits false signals. For example, an adversary is performing relay attack by modifying a message's timestamp and broadcast it several times in order to obstruct traffic [14, 15]. The IoV communicates with multiple infrastructures for its operation. The versatile nature of this communication opens the door for eavesdropping. [16, 17] explore the possibility of eavesdropping in roadside units (RSUs) and its counter measure using blockchain technology. They state that the impact of the adversary can get access to victim's social interactions on the internet as well as driver behavior. [18] states that using a transponder, a hacker can eavesdrop on communications between vehicles and RSUs and later sell the personal data of the vehicles, drivers, and passengers—including their names, contact information, social media handles, and photos—to advertising companies in order to generate revenue. Fog cloud based (IoV) computing enables the vehicular cloud to interact with the Internet and vice versa, Although, this cop-operation enhances the effectiveness of the IoV. However, [19] discusses the possibility of ED in cloud computing. [20, 21] discuss the possibility of eavesdropping in the vehicular cloud (VC). They state that the data can be eavesdropped on whenever a transmission is made with in the VC.

7.2.1.2 *Identity Exposure*

It is possible to track the route taken by a vehicle that has been the target of an observer and use the data for other purposes. A passive attacker can use the IoV system to reveal a target's identity. In addition to an ID, time, location, travel information, and other types of private data might be revealed [13, 22–24]. [25] discusses the issue of location theft in the cloud-based Internet of Vehicles. Identity theft can be done when a vehicle migrates from one virtual machine to another for better location services. [19] highlights the issue of identity exposure from the data stored in the cloud. [26] highlights that the ID can be exposed by the data stored on servers and preferences that are used by systems like infotainment.

7.2.1.3 *Traffic Analysis*

This is an attack which helps an adversary analyze communication even if it is encrypted. The adversary captures transmission patterns, number of packets delivered, and their timing between the vehicle and other things in the IoV. This will help the adversary find the relationship between the IoV and the other device [27, 28]. [14] states that the lack of centralized monitoring of network traffic can result in traffic analysis attacks.

Table 7.2 highlights the types of attacks on confidentiality in the IoV and their categories, reference papers, and purposes.

7.2.2 Attacks on Integrity of IoV Communication

Integrity deals with the correctness and accuracy of data. Due to this check, we are able to find if the transmitted data are modified. Hashing is a way of securing data integrity. Table 7.3 highlights the security threats to the integrity of IoV communication.

7.2.2.1 *Black Hole Attack*

In this attack, an adversary stops participating in network communication to form a black hole. Then the attacker tampers with the routing protocol by posing it as the shortest path to destination node [30]. This tricks other nodes, and they start transmitting data to the adversary. The adversary can try to interrupt network packets to either insert false information or cause the packets to be lost. This attack generates two possibilities that either result in denial-of-service (DoS) attacks or result in man-in-the-middle attack [31].

7.2.2.2 *Sinkhole Attack*

In this attack, an adversary routes the traffic through it. It aims to modify data contents before they are transmitted to a destination node [32]. Grey hole and black hole attacks, for example, can be mounted using this attack [33].

Table 7.2 Types of Attack, Categories of Attack, Reference Papers, and Purposes

Attack Type	*Category of Attack*	*Reference Papers*	*Purpose*
Eavesdropping	Active Attack	[14–21]	To capture the data of IoV communication.
Identity Exposure	Passive Attack	[13, 19, 22–26, 29]	To monitor the location of IoV communication.
Traffic Analysis	Passive Attack	[7, 14, 27, 28]	To monitor traffic patterns of IoV communication.

Table 7.3 Attacks on the Integrity of the IoV

Motive of Attack	*Types of Attacks*	*Weakness of Attacks*
To change the contents of data. This attack can pose serious threats like accidents and misleading vehicles, which in turn can cost precious lives and cause infrastructure damage.	• Black hole • Sinkhole • Gray hole • Message tampering • Masquerading • Malware	Standard symmetric encryption like AES-128Bits can avoid these types of attacks.

7.2.2.3 *Gray Hole Attack*

This works on the same concept as the black hole attack. However, instead of denial of service, this attack aims to degrade network performance by selectively dropping a particular type of message, like requests for prediction of energy consumption. In extreme cases, messages regarding road safety can be dropped [10]. This kind of attack is challenging to detect, as the adversary is acting intermittently [9].

7.2.2.4. *Message Tampering*

In this attack, the adversary intends to change the contents of the message to send wrong information [27]. This can result in creating chaos in the network. For example, an adversary node can transmit that a traffic jam or accident has occurred on the road, which in turn can divert the traffic to another path or call emergency services [34].

7.2.2.5 *Masquerading*

The adversary acts as a legitimate node. It receives messages and then sends false network communications or modifies the content of messages that were received. For instance, the attacker might learn that the route is clear from vehicles in front of it, yet it might broadcast the incorrect information as an emergency message, causing vehicles to slow down. This can result in a traffic jam [9].

7.2.2.6 *Malware*

This attack uses malicious software that is implanted in vehicles or in the infrastructure of the network. When this malicious software is activated, it can take down the whole network or even damage it [35]. [36] states that the attacker can get access to the vehicle for planting malware through various communication technologies like Wi-Fi, Bluetooth, 802.11p, LTE-V2X, and 5G (to name a few). Various Internet of Things (IoT) components can be infected using IoT botnets [37].

Table 7.4 highlights the types of attacks on integrity in the IoV and their reference papers and purposes.

7.2.3 Attacks on Availability of IoV Communication

Availability is responsible for keeping IoV resources available for authorized use either during normal operation or when the network is under attack by an adversary. Its goal is to ensure all the services of the network will be available in any circumstances. However, an adversary can hamper availability and enact the type of attack that can take place in the IoV environment with respect to availability. Table 7.5 summarizes attacks on the availability of the IoV.

Table 7.4 Attacks on Integrity of IoV, Their Categories, Reference Papers, and Purposes

Type of Attack	*Category of Attack*	*Reference Paper*	*Purpose*
Black hole	Active attack	[30, 31, 38–41]	Adversary takes control of the flow of messages.
Sinkhole	Active attack	[6, 9, 32, 33, 42–44]	Adversary modifies message contents.
Gray hole	Active attack	[6, 9, 10, 45–49]	Adversary degrades network performance.
Message tampering	Active attack	[27, 34, 50–57]	Adversary changes the contents of messages.
Masquerading	Active attack	[9, 58–65]	Adversary performs malicious activity in the network while acting as a legitimate node.
Malware	Active attack	[35–37, 45, 58, 66–70]	Malicious software performs the activity when it is activated.

Table 7.5 Attacks on the Availability of the IoV

Motive of Attack	*Types of Attacks*	*Weakness of Attacks*
To make network resources unavailable for legitimate users/ vehicles.	• Denial of service (DOS/DDOS) • Channel jamming • Spamming	If a malicious node can be identified, then its transmissions can be filtered out.

7.2.3.1 Denial of Service or Distributed Denial of Service

This is a threat in which an adversary makes network resources unavailable to legitimate users by overwhelming a service provider by sending it a huge number of requests. These requests saturate the service provider, which cannot process requests from legitimate users [69]. In an IoV environment, the connectivity is like a backbone which helps in making critical decisions like auto braking or increasing speed. If a timely decision cannot be made due to DoS/distributed denial of service (DDoS), it can cause loss of life or damage to vehicles (or other damage) [71]. [72] discusses DDoS in 5G networks and its mitigation.

7.2.3.2 Channel Jamming

It is done by sending high-frequency jamming signals either among the nodes in a cluster or among nodes and the infrastructure. [73–75] state that

Table 7.6 Attacks on Integrity of IoV and Their Categories, Reference Papers, and Purposes

Attack Type	*Category of Attack*	*Reference Papers*	*Purpose*
Denial of Service (DoS/DDoS)	Active Attack	[9, 69, 71, 72, 79, 80]	To make network resources either degraded or unavailable.
Channel Jamming	Active Attack	[6, 9, 31, 69, 73–75, 80]	
Spamming	Active Attack	[9, 49, 76–78, 81]	

there is a lack of consideration from a channel jamming point of view in the IoV infrastructure like roadside units (RSUs), GPS, and the cloud.

7.2.3.3 Spamming Attack

It is carried out by sending annoying messages to legitimate users. The adversary aims to consume network bandwidth to create delays in the network. [76] states that when the security of a vehicle is breached and an adversary takes the advantage of easily accessible information about the driver's life style, social relationships, and political convictions, it exposes the driver to spam. [77] states that the attacker can use machine learning algorithms for exploiting vehicles and personal vital information by targeting them with spamming messages. [78] states that spamming can be carried out in the unmanned aerial vehicle (UAV)–based IoV. Table 7.6 highlights attacks on the integrity of IoV and their categories, reference papers and purposes.

7.3 SOLUTIONS TO SECURITY THREATS IN THE IoV

In this section, we present remedies for different types of attacks in tabular form to make it convenient and easy to understand.

7.3.1 Counter Measures for Confidentiality of Information

This section contains discussion on counter measures of attacks on confidentiality of information. Figure 7.2 highlights attacks on confidentiality of the CR-IoV.

7.3.1.1 Eavesdropping

In order to mitigate eavesdropping, [19] proposed the authenticated key management protocol, which establishes session keys for secure communications after mutual authentication between communicating IoV entities. An NS2 simulator is used to implement the proposed method. [21]

proposed an efficient signcryption scheme for IoVs with an environment for certificateless cryptography (CLC) to send a message to the server in a public key infrastructure (PKI) environment. A random oracle model (ROM) formal analysis establishes the validity of the suggested methodology. Additionally, since there is no pairing involved in the computation, it is more efficient than the current protocols. A new mutual authentication and key agreement protocol in an IoV was put out by [15]. The suggested approach is also examined through formal and informal security analysis, formal security verification utilizing an automated verification tool, and security analysis in general. Low computational and communication overheads characterize this scheme. The suggested scheme's effectiveness is assessed using the NS2 simulator. [18] describes how accessing data may be authorized quickly in a reliable, safe, and decentralized way using blockchain technology to secure the CR-IoV. [16] evaluated the performance of proposed privacy management schemes CAPS, CPN, RSP, and SLOW. It concluded that SLOW is better than the others. In addition to this, a number of eavesdropping stations have a negative impact on the performance of these schemes. [82] highlights three counter measures against eavesdropping, SOADV, Ariadne, and SRP. SOADV uses digital signatures and one-way hashing to provide security. Ariadne provides security via one-way hashing. SRP establishes a secure connection between the source and destination using shared keys.

7.3.1.2 Identity Exposure

To overcome this attack, [15] proposed a novel mechanism for key exchange and mutual authentication in an IoV. The suggested approach is also examined through formal and informal security analysis, formal security verification utilizing an automated verification tool, and security analysis in general. Low computational and communication overheads characterize this scheme. The suggested scheme's effectiveness is assessed using the NS2 simulator. [23] proposed an IoV-SMAP after analyzing simulations from AVISPA and the real-or-random (ROR) model for internet security protocols and applications. It was determined that the IoV-SMAP offers superior security and effectiveness compared to rival methods. A secure message authentication protocol (SMEP-IoV) was suggested by [24] To address security demands, it employs compact hash algorithms and encryption techniques. The proposed SMEP-IoV is subjected to a rigorous security examination using BAN logic. According to the performance comparisons, the SMEP-IoV is quick and efficient, completing the authentication procedure in just 0.198 milliseconds. To increase the level of privacy protection while driving, [25] proposed a countermeasure against linkage mapping attacks during virtual machine (VM) migration in cloud-enabled Internet of Vehicles (CE-IoV). To counter this attack, a pseudonym-changing synchronization scheme (PCSS) is suggested. It is done by synchronization of the pseudonym-changing and VM identity-replacement processes to protect against attack. A concerted silence-based location privacy preserving scheme (CSLPPS) was proposed by [26]. In the IoV, CSLPPS ensures that users of IoV location-based services and vehicular networks are anonymous and untraceable. Cooperative automobiles simultaneously change their identities in this approach. Before starting their cyber-activity with the new identities, they observe a period of silence. The performance analysis and simulation demonstrated the effectiveness of CSLPPS against a simulated global passive attacker.

7.3.1.3 Traffic Analysis

In order to counter a traffic analysis attack, [7] proposed an anonymous key exchanging method. In this method, encryption of only vital data which can put privacy of the driver at risk is done.

7.3.2 Counter Measures on Integrity of Information

In this section, we will discuss counter measures that are used to defend against attacks on the integrity of information in the Internet of Vehicles. Figure 7.3 highlights attacks on the integrity of the CR-IoV.

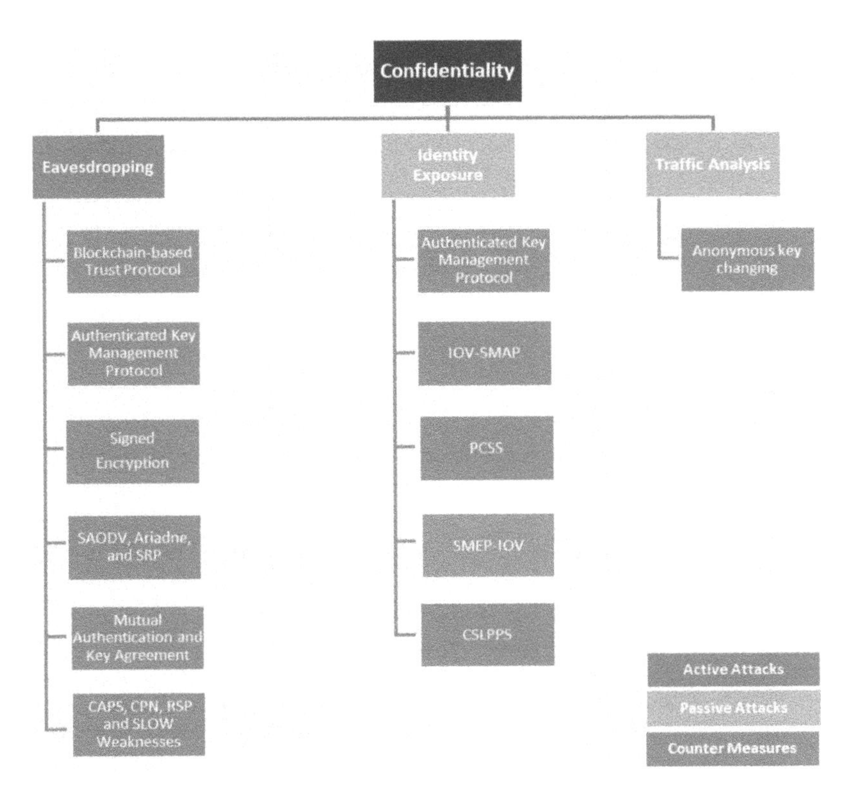

Figure 7.2 Attacks and counter measures on confidentiality in the CR-IoV.

7.3.2.1 Black Hole Attacks

[38] proposes the Rivest–Shamir–Adleman (RSA) algorithm to prevent the network from being attacked by a source node. It employs detection of sensor nodes or messages transmitted from sensor nodes to the base station and prevents network from being attacked by the source node. This ensures the security mechanism of the AODV protocol improves. MATLAB's simulation tool is used to run this simulation setup. In [39], by installing a monitoring system in each receiving node within the network, a widely used variable control chart is utilized to track node behavior and can quickly identify rogue nodes. This approach does not require changes to the routing protocol or the 802.11p standard. Utilizing a real map, the simulations include SUMO microscopic continuous road traffic simulators and NS-2 simulations. [30] proposed a delay-tolerant Internet of Vehicles with a decentralized reputation system called BiRep that identifies and penalizes black-hole nodes. Using the PRoPHET routing protocol, BiRep was tested. The results of the simulation demonstrated great performance in all circumstances, on a par with or better than existing reputation schemes, and a significant increase in the message delivery rate. [40] used secure weight-based AODV (SWAODV) with elliptic cryptography to mitigate black hole attacks.

7.3.2.2 Sinkhole Attacks

In [44], a blockchain-based solution is discussed to mitigate sinkhole attacks. Blockchain eliminates the middleman between individuals who do not find each other reliable by acting as a database for keeping a mutually trusted, joint, tamper-proof, timestamped count of records. The workload in a blockchain network is distributed over several computational devices rather than a central node. [42] highlighted the use of a parent failover and rank authentication scheme to counter sinkhole attacks.

7.3.2.3 Gray Hole Attacks

In [46], the strength of feature extraction offered by an autoencoder (AE) and the capacity of a support vector machine to utilize vast amounts of data were coupled by the authors. The experimental results demonstrate that the suggested model is highly accurate. [47] inquired into the application of nature-inspired algorithms in the routing and security of the IoV. They aimed to optimize all routing issues among vehicles, as delay in timely information cannot be tolerated in real-time applications. In addition to this, nature-inspired algorithms are able to prevent various attacks. [48] highlighted this use of SDGA; it maintains a record for each vehicle's activity on the network. A malicious node is identified using the Z-score. [49] employed deep packet inspection based on a convolutional neural network (CNN) classifier. To evaluate and test the effectiveness of the proposed system, a

WSN-DS dataset was used. The model achieved an accuracy of 97%. The proposed work can be used as a future benchmark for deep learning and intrusion prevention research communities in smart cities.

7.3.2.4 Message Tampering

In [52], the authors proposed a model using crowd-sensing for cloud-based nodes. In addition to this, location social infection theory is used to predict the vehicle location. Software for simulating an operational network environment is used to conduct experiments. The results of the experiments demonstrate that the proposed strategy is more competitive in resisting attacks when there are bad and selfish nodes in the Internet of Vehicles. In [53], the authors used the transmission of communications across automobiles using satellite communication and elliptic curve cryptography (ECC) for key agreement. The suggested protocol outperforms prior work in terms of security by using fewer communication bits for computations in the security algorithm and encrypting data at a moderately fast rate. [54] proposed a lightweight mechanism called DRiVe. It ensures message security as well as detecting malicious roadside units. Based on a probabilistic approach, DRiVe can detect malicious RSUs. Only when a vehicle enters or exits one RSU's coverage region do the authentication parameters get communicated. There are no computationally expensive cryptographic primitives used in DRiVe. By doing this, the security overhead of message authentication codes (MACs) with each packet is greatly reduced. The proposed method results in a 7.5% reduction in latency and a 7% reduction in the number of bits transferred. The proposed technique yields a probability of detection close to 99% for the case where hostile cars are present. [34] presents an authentication and secure data transfer algorithm in the IoV framework blockchain technology. It ensures secure communication via blockchain.

In order to protect information, [57] suggests an authentication and key exchange (AKE) system that is physically secure. In order to ensure that the system is secure even if the user devices or sensors are compromised, a physical unclonable function (PUF) is introduced in this work. The thorough security research shows that there are no known threats that could affect the proposed protocol.

7.3.2.5 Masquerade Attacks

In [61], the authors propose an efficient authentication scheme over blockchain named EASBF. Blockchain technology, one-way hashing, and elliptic curve cryptography are all used in the proposed EASBF scheme. Security analysis employs both the AVISPA tool and the random oracle model. [63] uses a secure technique for large-scale IoV big data collecting. Before connecting to the network, vehicles must first register with the big data center.

Vehicles then connect to the big data center using a single sign-on mechanism and mutual authentication. The discussion and performance evaluation findings demonstrate the proposed secure mechanism's efficacy and security.

7.3.2.6 *Malware Attacks*

A privacy-preserving based secured framework for the Internet of Vehicles (P2SF-IoV) was presented by the authors. Utilizing two network datasets, it combines deep learning and blockchain technology. The proposed method used two datasets namely ToN-IoT and IoT-Botnet [58]. For identifying and learning typical or deviant behavior among the IoV [68]. Deep learning (DL) methods of data exploration are applied. In the face of dangers and attacks, it makes intelligent decisions on its own. [36] This paper studies propagation dynamics on the Internet of Vehicles establishes an IoV-SIRS prevention model. The proposed model calculates the threshold of the virus and the degree of an outbreak. Simulation experiments show that the IoV-SIRS model has a good inhibitory effect on malicious virus spread. In this paper, the authors suggest using the distributed infrastructure to gather and preserve reliable evidence using Trust-IoV. The integrity of the evidence is guaranteed by Trust-IoV, which also enables investigators to confirm the integrity of the evidence while conducting an inquiry. According to empirical findings in a simulated environment, Trust-IoV works with little overhead [35]. [37] provided a quick overview of the structure of IoT botnets, their fundamental style of operation, significant DDoS occurrences involving IoT botnets in recent memory, and the related exploited vulnerabilities. Additionally, it offers suggestions and solutions to reduce the cyber dangers associated with the IoT. In order for vehicles to share network resources with enhanced trust, dependability, and security using distributed access control systems, this article integrates blockchain technology into ad hoc vehicular networking [70].

7.3.3 Counter Measures on Availability Attacks

In this section, we will provide counter measures for attacks on the availability of information in the IoV. Figure 7.4 highlights attacks on the availability of the CR-IoV.

7.3.3.1 *Distributed Denial of Service/Denial of Service*

A three-tier architecture is proposed in [71]. In this case, the central authority (CA) cooperates with the roadside units and makes sure that only authorized cars take part in the communication process. The effectiveness of the suggested RVAC is assessed based on the vehicle choice, communication loss, latency, false positive rate, and detection time. In assessing performance, vehicle density and adversary ratio are taken into account. The suggested RVAC maintains a high reputed vehicle selection rate with

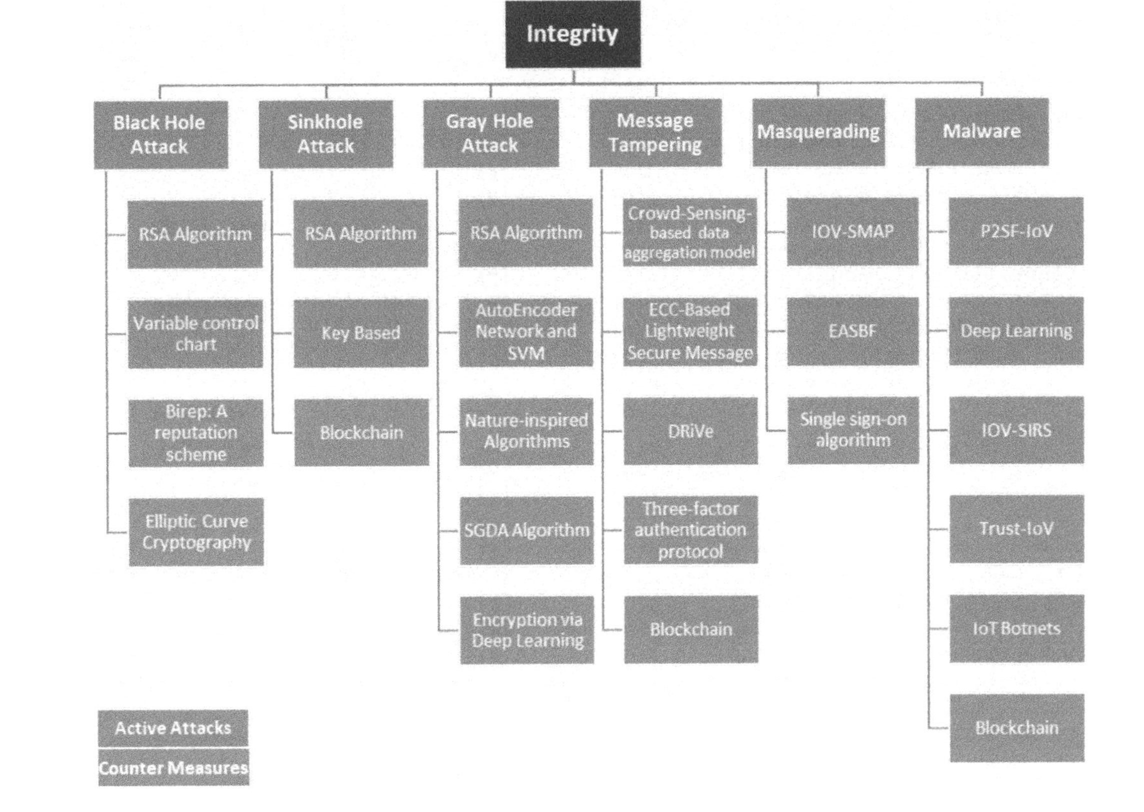

Figure 7.3 Attacks and counter measures on integrity of the CR-IoV.

Table 7.7 Various Threats and Their Counter Measures

Category	*Attack Type*	*Reference*	*Year*	*Proposed Solutions*
Confidentiality	**Eavesdropping**	[19]	2019	Authenticated Key Management Protocol
		[21]	2021	Signed Encryption
		[81]	2016	SAODV, Ariadne, and SRP
		[15]	2021	Mutual Authentication and Key Agreement Protocol
		[18]	2019	Blockchain
		[16]	2020	CAPS, CPN, RSP, and SLOW Weaknesses
	Identity Exposure	[15]	2019	Authenticated Key Management Protocol
		[23]	2020	IoV-SMAP
		[24]	2021	SMEP-IoV
		[25]	2021	PCSS
		[26]	2021	CSLPPS
	Traffic Analysis	[7]	2018	Anonymous Key Changing
Integrity	**Black Hole**	[38]	2021	RSA Algorithm
		[39]	2020	Variable Control Chart
		[30]	2021	Birep: A Reputation Scheme
		[40]	2021	Elliptic Curve Cryptography
	Sinkhole	[38]	2021	RSA Algorithm
		[42]	2019	Parent-Failover and Rank Authentication Scheme
		[44]	2020	Blockchain
	Gray Hole	[38]	2021	RSA Algorithm
		[46]	2022	Auto Encoder Network and Support Vector Machine
		[47]	2020	Nature-Inspired Algorithms
		[48]	2020	SDGA Algorithm
		[49]	2021	Encryption Scheme via Deep Learning

	Message Tampering	[52]	2020	Crowd-Sensing–Based Data Aggregation Model
		[53]	2020	ECC Based Lightweight Secure Message
		[54]	2021	DRiVe
		[57]	2017	Three-Factor Authentication Protocol
		[34]	2018	Blockchain
	Masquerading	[60]	2020	IoV-SMAP
		[61]	2021	EASBF
		[63]	2017	Mutual Authentication and Single Sign-On Algorithm
	Malware	[58]	2021	P2SF-IoV
		[68]	2020	Deep Learning
		[35]	2017	Trust-IoV
		[36]	2021	IoV-SIRS
		[37]	2017	IoT Botnets
		[70]	2019	Blockchain
Availability	**Denial of Service (DoS/DDoS)**	[71]	2018	Three-Tier Architecture
		[72]	2019	Virtualization-Based 5G Networks Using Management and Orchestration
		[80]	2020	RTED-SD: A Real-Time Edge Detection Scheme for Sybil DDoS
		[83]	2019	Enhanced Attacked Packet Detection Algorithm (EAPDA) as a Security Solution for DoS Attack
	Channel Jamming	[73]	2019	Delimited Anti-Jammer Scheme
		[74]	2019	PHY-Layer Cover-Free Coding for Wireless Pilot Authentication
		[75]	2020	Cooperative Relay Beam-Forming
		[80]	2020	RTED-SD: A Real-Time Edge Detection Scheme for Sybil DDoS
	Spamming	[77]	2021	ML-Based Spam Detection Scheme
		[81]	2022	Genetic Algorithm

reduced false positives, less communication loss, and shorter communication delay and detection time, according to experimental results. In [72], the authors suggested an approach which employs virtual machines (VMs) from intrusion prevention systems (IPSs) to intercept queries. The IPS's virtual machines are dynamically deployed via management and orchestration (MANO) in order to balance the load based on the amount of DDoS traffic. Experiments are carried out in a genuine 5G NFV environment that was developed utilizing 5G NFV environment tools to assess the effectiveness of the mechanism. The testing outcomes demonstrate that the suggested approach can successfully reduce DDoS attacks. A real-time edge detection scheme for Sybil DDoS in IoV was introduced by the authors in [80]. They quantify the traffic distribution using entropy theory and then create the fast quartile deviation check (FQDC) technique to detect and pinpoint DDoS attacks. To adapt the sliding window to the IoV environment, entropy values are incrementally calculated and optimized. All of the Sybil DDoS attacks reported in the F2MD datasets were successfully detected by the proposed model, which also has an average warning delay of 4.9193 seconds and an average TFOR of 1.6024%. In [83], the enhanced attacked packet detection algorithm prevents network performance from declining even in the face of this assault. In addition to verifying the nodes and identifying rogue nodes, EAPDA also increases throughput while reducing delay, hence boosting security. The simulator NS2 is employed.

7.3.3.2 Channel Jamming

For automobile traffic settings, the authors presented a machine learning-oriented delimited anti-jamming protocol in [73]. It focuses on differentiated signal recognition and filtration of jamming vehicles in order to pinpoint their exact location. To investigate the frequency fluctuations in signal strength brought on by jamming or other external attacks, a foster rationalizer is used. To forecast the locations of jamming vehicles, a decision tree–based machine learning open-sourced system called CatBoost is used. The proposed approach provides strong metrics for recall, F1 score, precision, and delivery accuracy. For multi-antenna V2I OFDM communications, a secure WPA (SWPA) protocol is constructed using PHY-layer cover-free (PHY-CF) coding theory [74]. Using cover-free coding, vehicle pilot signals are first encoded and then broadcast as a variety of subcarrier activation patterns (SAPs) in the time-frequency domain. To further achieve ultra-security, it uses pilot identification error probability (IEP) and demonstrates how the PHY-CF coding position may be used to pinpoint the attacker's location and lower IEP. To address control channel jamming issues in vehicle networks, a cooperative anti-jamming beamforming system was presented [75]. The multi-antenna RSU's spatial variety is used in this approach to relay cars and enhance control message transmission. The simulation results demonstrate that the suggested technique converges quickly, and when compared to benchmark systems, sizable performance advantages are discernible. In

[80], the authors proposed temporal false omission rate (TFOR) method. In this method, performance of reaction speed and omission rate is measured using a real-time edge detection scheme. All of the Sybil DDoS attacks supplied in the F2MD datasets were detected by the proposed method, which also had an average alert delay of 4.9193 seconds and an average TFOR of 1.6024%. In [83], the authors used an enhanced attacked packet detection algorithm to prevent network performance from declining even in the face of this assault. The nodes are verified by EAPDA, which also finds malicious nodes. Additionally, it increases throughput while minimizing delay. The simulator NS2 is employed.

7.3.3.3 Malware

IoT spam detection using machine learning was suggested by the authors in [77]. Each of the five machine learning models computes a spam score by taking the refined input attributes into account. The proposed method is tested using the REFIT Smart Home dataset. [81] employs a strategy based on optimization to identify botnet assaults in IoT systems. Botnet detection is done using a genetic algorithm with dynamic thresholds.

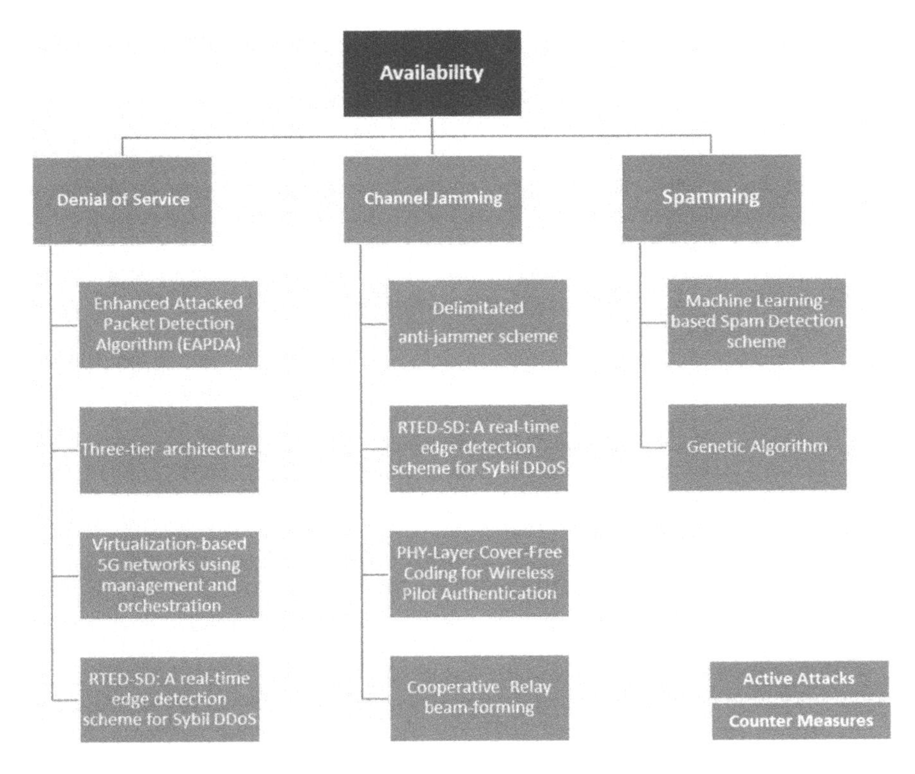

Figure 7.4 Attacks and counter measures for availability in the CR-IoV.

7.4 CONCLUSION

In this chapter, we have presented various security threats and their counter measures. In addition to this, we have also discussed them from three aspects, one from the aspect of network security goals, and we have categorized these attacks. In our work we have found that blockchain-based solutions are being used to counter various threats. One major potential work area that we have found is the use of machine learning and deep learning–based work to counter IoV security attacks.

REFERENCES

1. Y. Liu and G. Zhou, "Key technologies and applications of Internet of Things," *Proc. — 2012 5th Int. Conf. Intell. Comput. Technol. Autom. ICICTA 2012*, pp. 197–200, 2012, doi: 10.1109/ICICTA.2012.56.
2. R. P. Kumar, "Applications in internet of things (IoT)," *2018 2nd Int. Conf. Inven. Syst. Control*, no. ICISC, pp. 1156–1161, 2018.
3. Z. Mahmood, "Connected vehicles in the IoV: Concepts, technologies and architectures," *Connect. Veh. Internet Things Concepts, Technol. Fram. IoV*, pp. 3–18, Jan. 2020, doi: 10.1007/978-3-030-36167-9_1.
4. T. Alam, "A reliable communication framework and its use in Internet of Things (IoT)," *Int. J. Sci. Res. Comput. Sci. Eng. Inf. Technol.*, vol. 5, no. May, 2018, doi: 10.36227/techrxiv.12657158.
5. L. M. Ang, K. P. Seng, G. K. Ijemaru, and A. M. Zungeru, "Deployment of IoV for smart cities: Applications, architecture, and challenges," *IEEE Access*, vol. 7, no. December, pp. 6473–6492, 2019, doi: 10.1109/ACCESS.2018.2887076.
6. F. Sakiz and S. Sen, "A survey of attacks and detection mechanisms on intelligent transportation systems: VANETs and IoV," *Ad Hoc Netw.*, vol. 61, pp. 33–50, 2017, doi: 10.1016/j.adhoc.2017.03.006.
7. M. Abu Talib, S. Abbas, Q. Nasir, and M. F. Mowakeh, "Systematic literature review on internet-of-vehicles communication security," *Int. J. Distrib. Sens. Netw.*, vol. 14, no. 12, 2018, doi: 10.1177/1550147718815054.
8. L. Alouache, N. Nguyen, M. Aliouat, and R. Chelouah, "Survey on IoV routing protocols: Security and network architecture," *Int. J. Commun. Syst.*, vol. 32, no. 2, pp. 1–19, 2019, doi: 10.1002/dac.3849.
9. S. Sharma and B. Kaushik, "A survey on Internet of Vehicles: Applications, security issues & solutions," *Veh. Commun.*, vol. 20, p. 100182, 2019, doi: 10.1016/j.vehcom.2019.100182.
10. Y. Fraiji, L. Ben Azzouz, W. Trojet, and L. A. Saidane, "Cyber security issues of internet of electric vehicles," *IEEE Wirel. Commun. Netw. Conf. WCNC*, vol. 2018, no. April, pp. 1–6, 2018, doi: 10.1109/WCNC.2018.8377181.
11. N. Sharma, N. Chauhan, and N. Chand, "Security challenges in internet of vehicles (IoV) environment," *ICSCCC 2018–1st Int. Conf. Secur. Cyber Comput. Commun.*, pp. 203–207, 2018, doi: 10.1109/ICSCCC.2018.8703272.
12. J. Joy and M. Gerla, "Internet of vehicles and autonomous connected car—privacy and security issues," *2017 26th Int. Conf. Comput. Commun. Netw., ICCCN 2017*, pp. 0–8, 2017, doi: 10.1109/ICCCN.2017.8038391.

13. T. Garg, N. Kagalwalla, P. Churi, A. Pawar, and S. Deshmukh, "A survey on security and privacy issues in IoV," *Int. J. Electr. Comput. Eng.*, vol. 10, no. 5, pp. 5409–5419, 2020, doi: 10.11591/IJECE.V10I5.PP5409-5419.
14. Y. Sun *et al.*, "Attacks and countermeasures in the Internet of Vehicles," *Ann. des Telecommun.*, vol. 72, no. 5–6, pp. 283–295, 2017, doi: 10.1007/s12243-016-0551-6.
15. P. Bagga, A. K. Das, M. Wazid, J. J. P. C. Rodrigues, K. K. R. Choo, and Y. Park, "On the design of mutual authentication and key agreement protocol in internet of vehicles-enabled intelligent transportation system," *IEEE Trans. Veh. Technol.*, vol. 70, no. 2, pp. 1736–1751, 2021, doi: 10.1109/TVT.2021.3050614.
16. M. Babaghayou, N. Labraoui, A. A. A. Ari, M. A. Ferrag, and L. Maglaras, "The impact of the adversary's eavesdropping stations on the location privacy level in internet of vehicles," *SEEDA-CECNSM 2020–5th South-East Eur. Des. Autom. Comput. Eng. Comput. Netw. Soc. Media Conf.*, pp. 0–5, 2020, doi: 10.1109/SEEDA-CECNSM49515.2020.9221839.
17. U. Javaid, M. N. Aman, and B. Sikdar, "A scalable protocol for driving trust management in internet of vehicles with blockchain," *IEEE Internet Things J.*, vol. 7, no. 12, pp. 11815–11829, 2020, doi: 10.1109/JIOT.2020.3002711.
18. T. A. Butt, R. Iqbal, K. Salah, M. Aloqaily, and Y. Jararweh, "Privacy management in social internet of vehicles: Review, challenges and blockchain based solutions," *IEEE Access*, vol. 7, pp. 79694–79713, 2019, doi: 10.1109/ACCESS.2019.2922236.
19. M. Wazid, P. Bagga, A. K. Das, S. Shetty, J. J. P. C. Rodrigues, and Y. Park, "AKM-IoV: Authenticated key management protocol in Fog computing-based internet of vehicles deployment," *IEEE Internet Things J.*, vol. 6, no. 5, pp. 8804–8817, 2019, doi: 10.1109/JIOT.2019.2923611.
20. K. F. Hasan, T. Kaur, M. M. Hasan, and Y. Feng, "Cognitive Internet of Vehicles: Motivation, layered architecture and security issues," *2019 Int. Conf. Sustain. Technol. Ind. 4.0, STI 2019*, vol. 0, pp. 24–25, 2019, doi: 10.1109/STI47673.2019.9068070.
21. A. Elkhalil, J. Zhang, R. Elhabob, and N. Eltayieb, "An efficient signcryption of heterogeneous systems for internet of vehicles," *J. Syst. Archit.*, vol. 113, no. September, p. 101885, 2021, doi: 10.1016/j.sysarc.2020.101885.
22. M. S. Kim, "A survey of vehicular ad-hoc network security," *Lect. Notes Electr. Eng.*, vol. 425, pp. 315–326, 2018, doi: 10.1007/978-981-10-5281-1_35.
23. S. Yu, J. Lee, K. Park, A. K. Das, and Y. Park, "IoV-SMAP: Secure and efficient message authentication protocol for IoV in smart city environment," *IEEE Access*, vol. 8, pp. 167875–167886, 2020, doi: 10.1109/ACCESS.2020.3022778.
24. S. A. Chaudhry, "Designing an efficient and secure message exchange protocol for internet of vehicles," *Secur. Commun. Netw.*, vol. 2021, 2021, doi: 10.1155/2021/5554318.
25. D. Dujovne, T. Watteyne, X. Vilajosana, and P. Thubert, "6TiSCH : Deterministic IP-enabled industrial internet (of things)," *IEEE Commun. Mag.*, vol. 52, pp. 36–41, 2014. doi: 10.1109/MCOM.2014.6979984.
26. L. Benarous, S. Bitam, and A. Mellouk, "CSLPPS: Concerted silence-based location privacy preserving scheme for internet of vehicles," *IEEE Trans. Veh. Technol.*, vol. 70, no. 7, pp. 7153–7160, 2021, doi: 10.1109/TVT.2021.3088762.

27. D. B. Rawat, M. Garuba, L. Chen, and Q. Yang, "On the security of information dissemination in the internet-of-vehicles," *Tsinghua Sci. Technol.*, vol. 22, no. 4, pp. 437–445, 2017.
28. Hbaieb, Amal, Samiha Ayed, and Lamia Chaari. "A survey of trust management in the Internet of Vehicles." *Computer Networks*, vol. 203, p. 108558, 2022.
29. Y. Liu, Y. Wang, and G. Chang, "Efficient privacy-preserving dual authentication and key agreement scheme for secure V2V communications in an IoV paradigm," *IEEE Trans. Intell. Transp. Syst.*, vol. 18, no. 10, pp. 2740–2749, 2017, doi: 10.1109/TITS.2017.2657649.
30. C. Nabais, P. R. Pereira, and N. Magaia, "Birep: A reputation scheme to mitigate the effects of black-hole nodes in delay-tolerant Internet of Vehicles," *Sensors (Switzerland)*, vol. 21, no. 3, pp. 1–21, 2021, doi: 10.3390/s21030835.
31. B. Kwapong Osibo, C. Zhang, C. Xia, G. Zhao, and Z. Jin, "Security and privacy in 5G internet of vehicles (IoV) environment," *J. Internet Things*, vol. 3, no. 2, pp. 77–86, 2021, doi: 10.32604/jiot.2021.017943.
32. J. Arshad, M. A. Azad, R. Amad, K. Salah, M. Alazab, and R. Iqbal, "A review of performance, energy and privacy of intrusion detection systems for IoT," *Electronics*, vol. 9, no. 4, pp. 1–24, 2020, doi: 10.3390/electronics9040629.
33. L. Sleem, H. N. Noura, and R. Couturier, "Towards a secure ITS: Overview, challenges and solutions," *J. Inf. Secur. Appl.*, vol. 55, no. October, p. 102637, 2020, doi: 10.1016/j.jisa.2020.102637.
34. A. Arora and S. K. Yadav, "Block chain based security mechanism for internet of vehicles (IoV)," *SSRN Electron. J.*, Apr. 2018, doi: 10.2139/SSRN.3166721.
35. M. Hossain, R. Hasan, and S. Zawoad, "Trust-IoV: A trustworthy forensic investigation framework for the Internet of Vehicles (IoV)," *Proc. — 2017 IEEE 2nd Int. Congr. Internet Things, ICIOT 2017*, no. July, pp. 25–32, 2017, doi: 10.1109/IEEE.ICIOT.2017.13.
36. P. Xie, C. Fu, X. Wang, T. Feng, and Y. Yan, "Malicious atack pevention model of internet of vehicles based on IoV-SIRS," *Int. J. Netw. Secur.*, vol. 23, no. 5, pp. 835–844, 2021, doi: 10.6633/IJNS.202109.
37. K. Angrishi, "Turning internet of things (IoT) into internet of vulnerabilities (IoV) : IoT botnets," Feb. 2017, Accessed: Jan. 02, 2022. [Online]. Available: https://arxiv.org/abs/1702.03681v1.
38. P. Shah and T. Kasbe, "Detecting sybil attack, black hole attack and DoS attack in VANET using RSA algorithm," *2021 Emerg. Trends Indust. 4.0 (ETI 4.0)*, pp. 1–7, Dec. 2021, doi: 10.1109/ETI4.051663.2021.9619414.
39. B. Cherkaoui, A. Beni-Hssane, and M. Erritali, "Variable control chart for detecting black hole attack in vehicular ad-hoc networks," *J. Ambient Intell. Humaniz. Comput.*, vol. 11, no. 11, pp. 5129–5138, 2020, doi: 10.1007/s12652-020-01825-2.
40. M. Shukla, B. K. Joshi, and U. Singh, "Mitigate wormhole attack and blackhole attack using elliptic curve cryptography in MANET," *Wirel. Pers. Commun.*, vol. 121, no. 1, pp. 503–526, 2021, doi: 10.1007/s11277-021-08647-1.
41. Chung, W. J., "Modeling and simulation of blackhole attack detection using multipath routing in WSN-based IoV," *Int. J. Eng. Res.*, vol. V10, no. 1, 2021, doi: 10.17577/ijertv10is010075.
42. H. H. R. Sherazi, R. Iqbal, F. Ahmad, Z. A. Khan, and M. H. Chaudary, "DDoS attack detection: A key enabler for sustainable communication in Internet of

Vehicles," *Sustain. Comput. Informatics Syst.*, vol. 23, pp. 13–20, 2019, doi: 10.1016/j.suscom.2019.05.002.
43. E. Džaferović, A. Sokol, A. A. Almisreb, and S. Mohd Norzeli, "DoS and DDoS vulnerability of IoT: A review," *Sustain. Eng. Innov.*, vol. 1, no. 1, pp. 43–48, 2019, doi: 10.37868/sei.v1i1.36.
44. S. Showkat and S. Qureshi, "Securing the Internet of Things using blockchain," *Proc. Conflu. 2020–10th Int. Conf. Cloud Comput. Data Sci. Eng.*, pp. 540–545, 2020, doi: 10.1109/Confluence47617.2020.9058258.
45. M. Adhikari, A. Munusamy, A. Hazra, V. G. Menon, V. Anavangot, and D. Puthal, "Security and privacy in edge-centric intelligent internet of vehicles: Issues and remedies," *IEEE Consum. Electron. Mag.*, 2021, doi: 10.1109/MCE.2021.3116415.
46. S. Dadi and M. Abid, "Enhanced intrusion detection system based on autoencoder network and support vector machine," *Smart Innov. Syst. Technol.*, vol. 237, pp. 327–341, 2022, doi: 10.1007/978-981-16-3637-0_23.
47. S. Sharma and B. Kaushik, "A comprehensive review of nature-inspired algorithms for internet of vehicles," *2020 Int. Conf. Emerg. Smart Comput. Informatics, ESCI 2020*, pp. 336–340, 2020, doi: 10.1109/ESCI48226.2020.9167513.
48. K. Stepien and A. Poniszewska-Maranda, "Analysis of security methods in vehicular Ad-Hoc network against worm hole and Gray hole attacks," *Proc. — 2020 IEEE Int. Symp. Parallel Distrib. Process. with Appl. 2020 IEEE Int. Conf. Big Data Cloud Comput. 2020 IEEE Int. Symp. Soc. Comput. Netw. 2020 IEE*, pp.371–378,2020,doi:10.1109/ISPA-BDCloud-SocialCom-SustainCom51426.2020.00072.
49. D. Choudhary and R. Pahuja, "Encryption techniques for intelligent transportation systems via deep learning for IoV in smart cities," pp. 1–5, 2021, [Online]. Available: https://doi.org/10.21203/rs.3.rs-319815/v1.
50. A. Wahab, O. Ahmad, M. Muhammad, and M. Ali, "A comprehensive analysis on the security threats and their countermeasures of IoT," *Int. J. Adv. Comput. Sci. Appl.*, vol. 8, no. 7, 2017, doi: 10.14569/ijacsa.2017.080768.
51. Muhammad Anwar Shahid, Arunita Jaekel, Christie Ezeife, Qasim Al-Ajmi, and Ikjot Saini, *[IEEE 2018 Majan International Conference (MIC) – Muscat, Oman (2018.3.19-2018.3.20)] 2018 Majan International Conference (MIC) – Review of potential security attacks in VANET*, pp. 1–4, 2018, doi:10.1109/MINTC.2018.8363152.
52. W. Zhang and G. Li, "An efficient and secure data transmission mechanism for Internet of Vehicles considering privacy protection in Fog computing environment," *IEEE Access*, vol. 8, pp. 64461–64474, 2020, doi: 10.1109/ACCESS.2020.2983994.
53. C. T. Poomagal and G. A. Sathish Kumar, "ECC based lightweight secure message conveyance protocol for satellite communication in Internet of Vehicles (IoV)," *Wirel. Pers. Commun. 2020*, vol. 113, no. 2, pp. 1359–1377, May 2020, doi: 10.1007/S11277-020-07285-3.
54. N. V. Abhishek, M. N. Aman, T. J. Lim, and B. Sikdar, "DRiVe: Detecting malicious roadside units in the internet of vehicles with low latency data integrity," *IEEE Internet Things J.*, 2021, doi: 10.1109/JIOT.2021.3097809.
55. Y. Liu, Y. Wang, and G. Chang, "Efficient privacy-preserving dual authentication and key agreement scheme for secure V2V communications in an IoV

paradigm," *IEEE Trans. Intell. Transp. Syst.*, vol. 18, no. 10, pp. 2740–2749, Oct. 2017, doi: 10.1109/TITS.2017.2657649.

56. F. Sakiz and S. Sen, "A survey of attacks and detection mechanisms on intelligent transportation systems: VANETs and IoV," *Ad Hoc Netw.*, vol. 61, pp. 33–50, Jun. 2017, doi: 10.1016/J.ADHOC.2017.03.006.
57. Q. Jiang, X. Zhang, N. Zhang, Y. Tian, X. Ma, and J. Ma, "Three-factor authentication protocol using physical unclonable function for IoV," *Comput. Commun.*, vol. 173, pp. 45–55, May 2021, doi: 10.1016/J.COMCOM.2021.03.022.
58. R. Kumar, P. Kumar, R. Tripathi, G. P. Gupta, and N. Kumar, "P2SF-IoV: A privacy-preservation-based secured framework for internet of vehicles," *IEEE Trans. Intell. Transp. Syst.*, pp. 1–12, Aug. 2021, doi: 10.1109/TITS.2021.3102581.
59. S. Sharma and B. Kaushik, "A survey on Internet of Vehicles: Applications, security issues & solutions," *Veh. Commun.*, vol. 20, p. 100182, Dec. 2019, doi: 10.1016/J.VEHCOM.2019.100182.
60. S. Yu, J. Lee, K. Park, A. K. Das, and Y. Park, "IoV-SMAP: Secure and efficient message authentication protocol for IoV in smart city environment," *IEEE Access*, vol. 8, pp. 167875–167886, 2020, doi: 10.1109/ACCESS.2020.3022778.
61. M. S. Eddine, M. A. Ferrag, O. Friha, and L. Maglaras, "EASBF: An efficient authentication scheme over blockchain for fog computing-enabled Internet of Vehicles," *J. Inf. Secur. Appl.*, vol. 59, p. 102802, Jun. 2021, doi: 10.1016/J.JISA.2021.102802.
62. L. Yadav, S. Kumar, A. Kumarsagar, and S. Sahana, "Architechture, applications and security for IoV: A survey," *Proc.—IEEE 2018 Int. Conf. Adv. Comput. Commun. Control Networking, ICACCCN 2018*, pp. 383–390, Oct. 2018, doi: 10.1109/ICACCCN.2018.8748363.
63. L. Guo *et al.*, "A secure mechanism for big data collection in large scale internet of vehicle," *IEEE Internet Things J.*, vol. 4, no. 2, pp. 601–610, Apr. 2017, doi: 10.1109/JIOT.2017.2686451.
64. S. Tbatou, A. Ramrami, and Y. Tabii, "Security of communications in connected cars modeling and safety assessment," *ACM Int. Conf. Proceeding Ser.*, vol. Part F129474, Mar. 2017, doi: 10.1145/3090354.3090412.
65. N. Sharma, N. Chauhan, and N. Chand, "Security challenges in internet of vehicles (IoV) environment," *ICSCCC 2018–1st Int. Conf. Secur. Cyber Comput. Commun.*, pp. 203–207, Jul. 2018, doi: 10.1109/ICSCCC.2018.8703272.
66. S. Kim and R. Shrestha, "Internet of vehicles, vehicular social networks, and cybersecurity," *Automot. Cyber Secur.*, pp. 149–181, 2020, doi: 10.1007/978-981-15-8053-6_7.
67. E. S. Ali *et al.*, "Machine learning technologies for secure vehicular communication in internet of vehicles: Recent advances and applications," *Secur. Commun. Netw.*, vol. 2021, 2021, doi: 10.1155/2021/8868355.
68. M. S. H. Sassi and L. C. Fourati, "Investigation on deep learning methods for privacy and security challenges of cognitive IoV," *2020 Int. Wirel. Commun. Mob. Comput. IWCMC 2020*, pp. 714–720, Jun. 2020, doi: 10.1109/IWCMC48107.2020.9148417.
69. S. Abdus, A. Shadab, S. Mohammed, and B. Mohammad Ubaidullah, "Internet of vehicles (IoV) requirements, attacks and countermeasures," *5th Int. Conf. "Co mputing Sustain. Glob. Dev.*, no. March, pp. 4037–4040, 2018.

70. S. Sharma, K. K. Ghanshala, and S. Mohan, “Blockchain-based Internet of Vehicles (IoV): An efficient secure ad hoc vehicular networking architecture,” *IEEE 5G World Forum, 5GWF 2019 — Conf. Proc.*, pp. 452–457, Sep. 2019, doi: 10.1109/5GWF.2019.8911664.
71. A. Tolba and A. Altameem, “A three-tier architecture for securing IoV communications using vehicular dependencies,” *IEEE Access*, vol. 7, pp. 61331–61341, 2019, doi: 10.1109/ACCESS.2019.2903597.
72. S. Köksal, Y. Dalveren, B. Maiga, and A. Kara, “Distributed denial-of-service attack mitigation in network functions virtualization-based 5G networks using management and orchestration,” *Int. J. Commun. Syst.*, vol. 34, no. 9, pp. 1–16, 2021, doi: 10.1002/dac.4825.
73. S. Kumar, K. Singh, S. Kumar, O. Kaiwartya, Y. Cao, and H. Zhou, “Delimitated anti jammer scheme for Internet of Vehicle: Machine learning based security approach,” *IEEE Access*, vol. 7, pp. 113311–113323, 2019, doi: 10.1109/ACCESS.2019.2934632.
74. D. Xu, P. Ren, and J. A. Ritcey, “PHY-layer cover-free coding for wireless pilot authentication in IoV communications: Protocol design and ultra-security proof,” *IEEE Internet Things J.*, vol. 6, no. 1, pp. 171–187, 2019, doi: 10.1109/JIOT.2018.2878333.
75. P. Gu, C. Hua, W. Xu, R. Khatoun, Y. Wu, and A. Serhrouchni, “Control channel anti-jamming in vehicular networks via cooperative relay beamforming,” *IEEE Internet Things J.*, vol. 7, no. 6, pp. 5064–5077, 2020, doi: 10.1109/JIOT.2020.2973753.
76. P. Zhao, G. Zhang, S. Wan, G. Liu, and T. Umer, “A survey of local differential privacy for securing Internet of Vehicles,” *J. Supercomput.*, vol. 76, no. 11, pp. 8391–8412, 2020, doi: 10.1007/s11227-019-03104-0.
77. A. Makkar, S. Garg, N. Kumar, M. S. Hossain, A. Ghoneim, and M. Alrashoud, “An efficient spam detection technique for IoT devices using machine learning,” *IEEE Trans. Ind. Informatics*, vol. 17, no. 2, pp. 903–912, 2021, doi: 10.1109/TII.2020.2968927.
78. K. N. Qureshi, M. A. S. Sandila, I. T. Javed, T. Margaria, and L. Aslam, “Authentication scheme for unmanned aerial vehicles based internet of vehicles networks,” *Egypt. Informatics J.*, no. xxxx, 2021, doi: 10.1016/j.eij.2021.07.001.
79. S. Sharma, K. K. Ghanshala, and S. Mohan, “A security system using deep learning approach for internet of vehicles (IoV),” *2018 9th IEEE Annu. Ubiquitous Comput. Electron. Mob. Commun. Conf. UEMCON 2018*, pp. 1–5, 2018, doi: 10.1109/UEMCON.2018.8796664.
80. J. Li, Z. Xue, C. Li, and M. Liu, “RTED-SD: A real-time edge detection scheme for sybil DDoS in the Internet of Vehicles,” *IEEE Access*, vol. 9, pp. 11296–11305, 2021, doi: 10.1109/ACCESS.2021.3049830.
81. S. Rethinavalli and R. Gopinath, “Botnet attack detection in internet of things using optimization techniques,” *Int. J. Electr. Eng. Technol.*, vol. 11, no. 10, pp. 412–420, 2022, doi: 10.34218/IJEET.11.10.2020.052.
82. Y. Sun *et al.*, “Attacks and countermeasures in the Internet of Vehicles,” *Ann. des Telecommun.*, vol. 72, no. 5–6, pp. 283–295, 2017, doi: 10.1007/s12243-016-0551-6.
83. A. Singh and P. Sharma, “A novel mechanism for detecting DOS attack in VANET using enhanced attacked packet detection algorithm (EAPDA),” *2015 2nd Int. Conf. Recent Adv. Eng. Comput. Sci. RAECS 2015*, no. December, pp. 1–5, 2016, doi: 10.1109/RAECS.2015.7453358.

Chapter 8

QoS Provisioning in CR-Based IoV

Muhammad Zeeshan

Intelligent transportation systems (ITSs) are expected to largely benefit from the recent technological advancements in the Internet of Things (IoT). The inclusion of the IoT in vehicular networks will certainly result in the improvement of road safety and reduction of road accidents. The Internet of Vehicles (IoV) is a special case of the IoT, as it defines a global network of vehicles and communicating units empowered with advanced wireless communication technologies and protocols [1]. These communication protocols help interconnect the vehicles on the road and also provide connectivity among various system units. The basic object in an IoV paradigm is a vehicle that is technically considered a smart multi-sensor entity equipped with various computational units, IP-based connectivity, and various communication technologies. Although the term IoV is sometimes used interchangeably with vehicular ad hoc network (VANET), it has certain extended features and capabilities. Broadly speaking, an IoV may be considered a business-oriented architecture, encompassing internet and multiple heterogeneous wireless access networks. These networks usually have very large target domains including road safety, infotainment, traffic management, and optimization.

To support communication in vehicular networking and the IoV, various spectrum regulatory authorities have allocated designated bands to enable vehicular communication. For instance, the Federal Communications Commission (FCC) has officially allocated a 75-MHz wide spectrum in the 5.9 GHz band as a licensed spectrum for vehicular communication. With an immense increase in the number of nodes in IoV, the officially allocated spectrum is not always available. Therefore, there is a need to apply cognitive radio (CR) technology to enhance spectrum utilization by using opportunistic spectrum access [2]. The induction of CR technology in IoVs will result in better spectrum utilization and provision of necessary cognition per the service requirements of various nodes and channel conditions. Despite these advantages, a key challenge in the realization of CR-based IoVs is heterogeneity in the quality-of-service (QoS) and data flow requirements imposed by different types of nodes present in the network.

Various IoV applications require different levels of service qualities under varying channel conditions, operating ranges, and mobility scenarios with

DOI: 10.1201/9781003284871-13

different constraints [3]. For instance, emergency information such as collision alert messages and security alarms is typically not data intensive but requires latency less than 100 ms. On the other hand, infotainment applications may tolerate the delay but have to support higher volumes of data. Provisioning of these diverse QoS requirements demands cognitive communication and data traffic scheduling mechanisms in the context of the CR regime. In this chapter, we will introduce various applications and scenarios for the IoV in context of their QoS requirements. We will then describe the problem of QoS provisioning in the context of CR-based networks and present cognitive communication mechanisms based on the service requirements of various vehicular nodes, considering link quality and mobility. Furthermore, we will also describe a QoS-aware data traffic scheduling framework for a prioritized flow of information.

8.1 QOS REQUIREMENTS OF IoV APPLICATIONS

With the increasing number of nodes in modern IoV scenarios, heterogeneity in the QoS requirements of various IoV applications is on the rise. From highly critical alarm messages to data-intensive infotainment applications, modern IoVs possess highly different service quality requirements in terms of latency, throughput, and reliability. At one end, there are delay-sensitive applications that are required to be serviced in real time. They have very strict latency and reliability requirements, especially if this information contains emergency or alarm messages, such as collision alerts. On the other end, there are data-intensive applications that can tolerate a latency up to hundreds of milliseconds, but the data being generated is tremendously large, requiring higher throughput and quick processing. Although the inclusion of CR technology in the IoV results in better spectrum utilization and provision of necessary cognition through opportunistic spectrum access, a key challenge is to meet the diverse QoS and data flow requirements under spectrum sensing uncertainties.

Before presenting the QoS requirements of various IoV applications, we briefly describe the potential use cases and typical scenarios encountered in vehicular communication.

8.1.1 Vehicle Platooning

Vehicle platooning is a concept of operating multiple vehicles in a closely connected mechanism, as shown in Figure 8.1. In this setting, vehicles move similarly to a train, with wireless connectivity among them. To realize this concept, they need to share status information, such as heading directions, speed, and braking/accelerating actions [4]. The advantage of this topology is that fuel consumption and the number of required drivers can be reduced by decreasing the distance between the vehicles. The communication aspects of this topology may include the intention to join or leave a platoon, intention to be a leader or follower, announcement or warning to non-member vehicles about an operational platoon, and messages for platoon management. The messages for platoon management include information about when to brake or accelerate, when to follow which path, when to change lanes, and requests to change the platoon leader (as the leading vehicle continuously consumes more fuel). Furthermore, these platoon messages should be secured to avoid any potential threats. The communication range in this topology is usually related to the length of complete platoon. To avoid communication interruption, the length of the platoon (i.e. number of vehicles) needs also to be managed. The potential QoS and data flow requirements of this use case are relatively relaxed. It requires supporting at least 30 messages per second periodically, with a variable message payload of 50–1200 bytes, excluding any security-related short messages. The latency requirement for this scenario is 10–25 ms, the throughput requirement is 12 Mbps, and the reliability demand is 90–99% [4].

8.1.2 Sensor and State Map Sharing

The purpose of sensor and state map sharing (SSMS) is to build joint situational awareness through sharing of raw or processed data acquired from various sensors [5]. This mode requires low-latency communication to realize the highly precise positioning and control that are necessary to empower mission-critical applications, such as vehicle platooning and cooperative driving to provide safety to all road users, including emergency vehicles and pedestrians. It is imperative that high-resolution images from sensors be communicated, resulting in higher bandwidth usage. To prevent this, smart nodes may also perform on-board processing of data for shared situational awareness, and tactical planning operations may be performed by the participating vehicles autonomously. Even with this improved processing by vehicles and nodes, significant bandwidth is required for SSMS. In addition, it requires high reliability for fusion confidence of situational awareness, low latency to allow emergency vehicle response, and large message payload size. The potential service quality requirements include message size of up

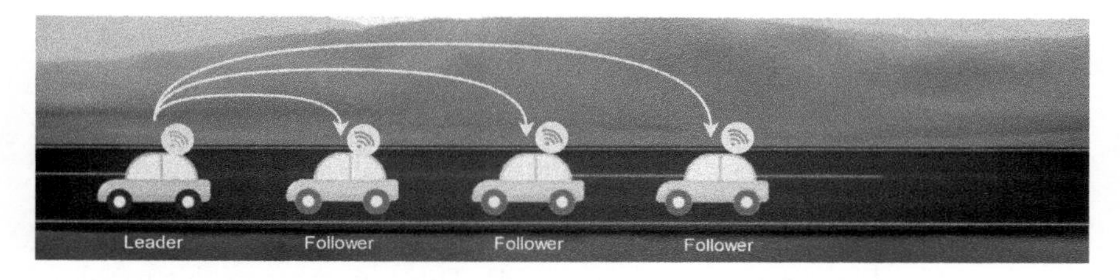

Figure 8.1 Concept of vehicle platooning with a leader vehicle connected to a group of follower vehicles.

to 600 bytes, latency between 3 and 100 ms, a data rate of 25 Mbps, and reliability of up to 99.999%.

8.1.3 Remote Driving

This mode allows a vehicle to be controlled remotely through cloud computing or a human operator. In the case of human operators, remote driving may be accomplished by using a lower number of sensors compared to fully autonomous control. For instance, a remote human operator can easily understand potential hazards using the live video feed from a vehicle's onboard camera and send necessary commands to the corresponding vehicle. The situation becomes more challenging when it comes to remotely controlling the vehicle by cloud computing through the use of many sensors and sophisticated algorithms without human interaction. It also requires coordination between vehicles to minimize the risk of traffic congestion, reduce overall travel time, and lead to enhanced fuel efficiency.

Due to the use of a large number of sensors and huge amount of aggregated data to be transmitted in the uplink, the throughput requirements of remote driving scenario are as high as 20 Mbps in the uplink. On the other hand, the downlink information contains command and control signals, resulting in a throughput requirement of up to 1 Mbps. The latency requirement is 5 ms due to the critical nature of information in both the uplink and downlink. Further, remote driving also requires very high reliability of at least 99.999% to avoid any misinformation that may result in collisions and other damage [4].

8.1.4 Advanced Driving

Advanced driving enables vehicles to perform better predictions of the probability of an accident through cooperative collision avoidance (CoCA). It involves the exchange of information among vehicles consisting of the sensors' data, safety messages, commands regarding braking and accelerating actions, and lateral/longitudinal control. For fully automated driving, high-resolution perception data is required to be shared among the nearby vehicles. In order to perform a cooperative maneuver, a detailed expected trajectory also needs to be shared among all the involved nodes. A critical set of information in advanced driving, referred to as the emergency trajectory alignment (EtrA), needs to be shared to help the driver in challenging and dangerous driving situations for traffic safety enhancement. In the case of unforeseen road and traffic conditions, EtrA messages consisting of the status information, sensor data, and information for coordinated maneuvers bring increased safety. Unexpected road conditions include sudden collisions on the road, loss of goods, animals crossing, and pedestrians on the road. In this case, a vehicle gathers the information from the on-board sensors and informs other vehicles about the unexpected and critical situation immediately. Such information requires very high reliability and ultra-low latency. Upon receiving these emergency messages, the nearby vehicles take necessary action by aligning their paths cooperatively.

This QoS requirements of this use case are relatively stringent compared to the other cases. The communication infrastructure should be able to support a message transmission rate of about 10–50 messages per second with packet sizes varying from 450 to 6500 bytes and maximum latency ranging from 3 to 100 ms. The variation of latency and message size requirements depends on the nature of the messages being transmitted. To ensure the transmission of accurate information in a timely manner, the reliability requirement of this use case is as high as 99.999%, with a throughput demand of 25 Mbps in the uplink.

Table 8.1 shows the summarized QoS requirements of some typical IoV scenarios in terms of data packet size, message transmission rate, maximum latency, reliability, and data rate. It is apparent that IoV scenarios have very diverse QoS requirements, giving rise to the need for appropriate data traffic scheduling and QoS provisioning frameworks.

8.2 PRIORITY-WISE CLASSIFICATION OF DATA

An in-depth overview of the QoS requirements of some key IoV scenarios is described in section 8.1. Each scenario has its own QoS requirements, including latency, throughput, and reliability. A key challenge in the QoS provisioning of CR-based IoVs is to incorporate all three indicators. It was mentioned earlier that QoS provisioning usually consists of two parts. The first part, called data traffic scheduling, refers to the priority-wise classification, data queue formation, and subsequent priority-aware scheduling of the data. Such scheduling approaches are mostly based on the demanded latency only. A better approach would be to incorporate the packet size and

Table 8.1 QoS Requirements of Some Important IoV Scenarios [**6**]

	Vehicle Platooning	*SSMS*	*Remote Driving*	*Advanced Driving*
Message size (bytes)	50–1200	1600	1000	450–6500
Message transmission rate (per second)	30	50	30	10–50
Maximum latency (ms)	10–25	3–100	5–20	3–100
Reliability	90–99%	99–99.999%	99.999%	99.99–99.999%
Throughput (Mbps)	12	25	1 (downlink), 25 (uplink)	0.5 (downlink), 25 (uplink)

Table 8.2 Sub-Classes Based on Latency Requirements

Sub-Class	*Priority*	*Latency Range (ms)*	*Relevant IoV Scenario(s)*
L_1	High	$3 \leq L \leq 10$	Remote driving, critical data in SSMS, and advanced driving
$L2$	Medium	$10 < L \leq 50$	Specific data in all scenarios
$L3$	Low	$50 < L \leq 100$	Advanced driving and SSMS

Table 8.3 Sub-Classes Based on Throughput Requirements

Sub-Class	*Category*	*Throughput Range (Mbps)*	*Relevant IoV Scenario(s)*
$D1$	High	$12 \leq D \leq 25$	Vehicle platooning, SSMS, and uplink of advanced and remote driving
$D2$	Low	$D \leq 1$	Downlink in advanced and remote driving

Table 8.4 Priority-Wise Categorization (Descending Order of Priority) Depending on the Nature of Data

Nature of Data	*Description*
IE	Interrupt emergency
E	Emergency
I	Interrupted
N	Normal

throughput requirements of each scenario to efficiently utilize the available spectrum in the case of a CR-based network. The second part, called the cognitive communication framework, focuses on the QoS-aware adaptive mechanism using the hybrid waveform concept based on the latency, reliability, and throughput requirements to efficiently utilize the available variable bandwidth spectrum holes. We will explain both these approaches in detail later in this chapter. In this section, we present a priority-aware classification of IoV data based on the latency and throughput requirements. It is worth mentioning that since reliability is a measure of the data integrity or QoS (e.g. bit or packet error rate), it is considered in the cognitive communication part (Section 8.4).

8.2.1 Latency, Throughput, and Data Type Considerations

We refer to each node in the CR-based IoV as a cognitive vehicular node (CVN), defined as a fundamental entity of the CR-based IoV network which is involved in two-way communication. Based on the latency and throughput requirements, we categorize all CVNs into two groups of sub-classes, similar to the strategy used in our previous work [7]. The first group is based on the latency (L) requirement, which is composed of three sub-classes, L_1 (highest priority), L_2 (medium priority), and L_3 (lowest priority). All these sub-classes, along with the range of demanded latency values, are shown in Table 8.2. The second group is based on throughput (D) requirements and is composed of two sub-classes, D_1 and D_2. Table 8.3 shows the second group of sub-classes, along with the throughput requirements and corresponding IoV scenarios. In addition to these QoS-based classifications, there is a need to classify the data of each CVN depending on the nature of the data. In other words, there are some special types of data that need to be treated uniquely in terms of priority, for instance, EtrA messages in advanced driving, emergency vehicle response in SSMS, security-related messages in vehicle platooning, and interrupted data (due to the arrival of a primary user (PU) or other higher-priority data) of any kind/scenario. In view of this, the CVN's data is categorized into various types, shown in Table 8.4 in descending order of priority. The first and highest priority is given to the data which contains very critical alarm messages or control actions, and it is interrupted due to the possible arrival of a PU. It is referred to as interrupt emergency (IE) data. The emergency data (E) that is scheduled for transmission but was not interrupted in the previous channel allocation has a lower priority compared to IE data. The third type is the usual data (non-emergent) of any priority class that was interrupted (I) and its transmission was previously incomplete. Here, the reason for interruption is due to the possible arrival of either a PU or other higher-priority data. Any other non-emergent and non-interrupted data belonging to any latency or throughput sub-class is categorized as normal data (N).

8.2.2 Priority Classes

The QoS-aware priority classes are formulated using the combination of both the latency and throughput sub-classes. The number of classes depends on the available variable bandwidth channels and number of sub-classes in the latency and throughput priority groups. The process can be made simpler by categorization of the available channels based on their bandwidth. It is intuitive to consider the bandwidth of available channels when assigning priority to given data to efficiently and purposely utilize the available spectrum. Based on the variation of bandwidth in the available channels as a result of spectrum sensing, multiple bandwidth categories may be devised. For simplicity, we assume that the available channels are either high or low bandwidth to support throughput ranges D_1 and D_2, respectively. For the CR-based IoV network, PUs will always have the highest priority, irrespective of the channel bandwidth. The priority classes are named C_i^k, where $k = 1$ for low bandwidth and $k = 2$ for high bandwidth, and $i = 0,1,\ldots,7$, where C_0^k has the highest priority and C_7^k has the lowest priority for the k_{th}

Table 8.5 Priority Classes (Descending Order) for Low- and High-Bandwidth Channels

Low Bandwidth $(\mathbf{k}=1)$		*High Bandwidth* $(\mathbf{k}=2)$	
Class	*Sub-Class Pair*	*Class*	*Sub-Class Pair*
C_0^1	PU	C_0^2	PU
C_1^1	(L_1, D_2)	C_1^2	(L_1, D_1)
C_2^1	(L_1, D_1)	C_2^2	(L_2, D_1)
C_3^1	(L_2, D_2)	C_3^2	(L_1, D_2)
C_4^1	(L_2, D_1)	C_4^2	(L_2, D_2)
C_5^1	(L_3, D_2)	C_5^2	(L_3, D_1)
C_6^1	(L_3, D_1)	C_6^2	(L_3, D_2)

bandwidth. Table 8.5 shows the complete details of these priority classes in descending order of priority. The priority assignment usually depends on how much weight is being given to latency and throughput while utilizing the available spectrum efficiently. As an example, we use a rule that higher priority will be given to more critical low-throughput data compared to low-critical high-throughput data if a low-bandwidth channel is available $(k=1)$. This is set to allow timely transmission of critical messages to avoid collisions and damage by making use of the instantly available low bandwidth channel(s). On the other hand, if a high-bandwidth channel is available $(k=2)$, then low-critical high-throughput data will be given higher priority compared to more critical low-throughput data. Here, the reason is to better utilize the available high-bandwidth channel while meeting the latency requirements of critical data.

8.3 TRANSMISSION SCHEDULING PROCESS

In this section, we present the priority-aware data traffic scheduling process, considering the priority classification of various types of data generated in CR-based IoV networks. In this regard, the first and foremost step is the formation of priority queues consisting of all the CVN data (depending on the scenario and nature of the data). This is followed by the transmission scheduling process that allocates suitable channel(s) to fulfil the different service requirements of each CVN by minimizing the overall system cost (to be discussed later in this section).

We assume that there is a fixed time interval (ΔT) for which a channel is allocated to a CVN already present in a queue. It is also assumed that a PU will always arrive at the start of a time slot to avoid interruption of the CVN's transmission during the slot. Consider a CR-based IoV scenario having a total of N CVNs. The current state of the nth CVN is defined as a vector that includes the number of packets to be transmitted in the mth slot (denoted by $\alpha_n(m)$) and the associated flags for emergency and interrupt data and is given by

$$s_n(m) = [\alpha_n(m), I_n(m), E_n(m)]. \quad (8.1)$$

The complete state vector of all the CVNs for the mth slot also needs to specify the channel occupancy status. For this, we define $\mathbf{c}_0(m)$ as the channel occupancy vector for the mth slot that is set to unity if a channel is available and zero otherwise. The length of this vector is equal to C, which is the total number of channels under consideration. The complete state vector, $\mathbf{s}(m)$, is given by

$$\mathbf{s}(m) = [\mathbf{c}_o(m), s_n(m) | n = 0,1,\ldots,N-1]. \quad (8.2)$$

With the complete state vector in hand, the channel allocation policy is now defined. Starting with the mth slot, the transmission scheduler will allocate an appropriate channel to each CVN according to a channel allocation policy $\gamma(m)$, a vector of length N for the mth slot, such that

$$\gamma_n(m) = \begin{cases} 0 & \text{no channel is assigned to } n^{th} \text{ CVN} \\ j & j^{th} \text{ channel is assigned to } n^{th} \text{ CVN} \end{cases} \quad (8.3)$$

Note that a channel can only be assigned to a CVN if it has non-empty packets to transmit. Continuing this process for all the CVNs with non-empty packets (i.e. $\alpha_n(m) > 0$) and the number of available channels in the mth time slot, a decision space $\Gamma(c_o(m), \alpha_n(m))$ is populated which includes all the possible permutations of decision policies.

Based on the nature of data and the priority classes defined in Tables 8.4 and 8.5, respectively, a total of five queues, q_{PU}, q_{IE}, q_E, q_I, and q_N, are maintained. Note that there will be separate priority queues for each channel bandwidth. In the scenario under consideration, there will be two sets of queues, low-bandwidth queues and high-bandwidth queues. The complete concept of the transmission scheduling process along with the two sets of priority queues is illustrated in Figure 8.2. An important thing to observe is that the emergency data only belongs to those classes that have very strict latency requirements, L_1, as their subclass. Therefore, if the nature of data is critical or emergency, there will only be two priority classes in the queue. In the transmission scheduling process, the primary users' queue (q_{PU}) has the highest priority. For all other users (i.e. CVNs), transmission priority decreases from q_{IE} to q_N. The transmission priority decreases vertically within each queue in accordance with the priority shown in Table 8.5. Since we have categorized the available channels into two bandwidth types, only one of the two sets of priority queues is selected based on the bandwidth of each available channel. This can be extended to more than two bandwidth

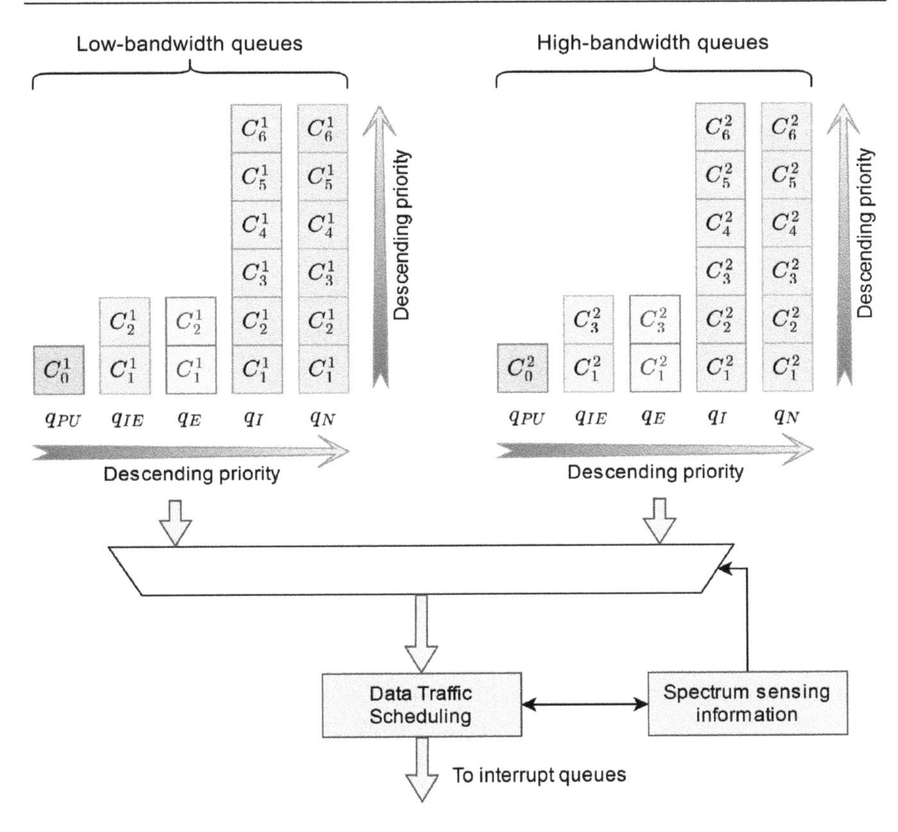

Figure 8.2 Priority-aware queues and data transmission scheduling in CR-based IoV [**8**].

types, depending upon the variation in the channel bandwidths as a result of various spectrum sensing approaches and operational scenarios. This traffic scheduling framework will allocate suitable channels to SCNs by using the channel allocation policy. The transmission scheduling process involves the assignment of suitable weights to each CVN class along with the nature of their data (i.e. IE, E, I, or N). Let W_i^k be the weight of priority class C_i^k and W_{IE}, W_E, and W_I be the weights corresponding to IE, E, and I data types, respectively. The values of these weights need to be selected to guarantee the proper functionality of the priority queues, as shown in Figure 8.2. The overall weight of the nth CVN is defined as

$$W_n = W_{i,n}^k + W_I \mathbf{1}_{(I_n=1)} + W_E \mathbf{1}_{(E_n=1)} + W_{IE} \mathbf{1}_{(I_n=1,E_n} \tag{8.4}$$

where $W_{i,n}^k$ is the weight assigned to the nth newly arrived CVN with data belonging to the ith priority class and k^{th} bandwidth, and $\mathbf{1}_{(x=1)}$ is the indicator function that is unity for $x = 1$ and zero otherwise.

Every time the nth CVN requests the transmission of its data, it is allocated a time slot, say, m, and $\alpha_n(m)$ is set to its corresponding number of packets. If applicable, the interrupt and/or emergency flags are also set. Assuming a total of C available channels, let L and H be the number of low- and high-bandwidth channels, respectively. For all existing CVNs, the number of packets is updated in each time slot, based on the channel allocation policy, as follows:

$$\alpha_n(m+1) = \begin{cases} \alpha_n(m), & \text{if } \gamma_n = 0 \\ \alpha_n(m) - P_L, & \text{if } \gamma_n = 1,2,\ldots,L \\ \alpha_n(m) - P_H, & \text{if } \gamma_n = L+1,\ldots,C \end{cases} \tag{8.5}$$

where P_L and P_H are the number of packets transmitted in a single time slot over low- and high-bandwidth channels, respectively, considering the same slot duration for each bandwidth.

One of the key aspects of QoS provisioning in CR-based IoVs is to minimize the overall latency and maximize the overall throughput of the complete network while achieving the desired service requirements. To achieve this end, it is necessary to formulate a utility function that can be optimized under certain QoS constraints. First, we will formulate individual utility functions for latency and throughput and then combine both of them into a single utility function. As stated earlier, the duration of each time slot is fixed to ΔT; therefore, the transmission delay experienced by the nth CVN will be ΔT if no channel is allocated to it in the mth slot. On the other hand, if a channel is allocated to the nth CVN in the mth slot, the transmission delay will be zero. Note that the transmission delay is an indication of how long a particular node needs to wait before a slot allocation. Therefore, the transmission delay of the nth CVN is given by

$$T_n(m) = \begin{cases} \Delta T, & \text{if no channel is allocated to } n^{th} \text{ CVN} \\ 0, & \text{Otherwise} \end{cases} \tag{8.6}$$

The overall latency corresponding to all CVNs, referred to as the latency utility function ρ_L, is defined as the weighted sum of the individual transmission delays of all CVNs with non-zero packets. Note that ρ_L is a function of the state vector $\mathbf{s}(m)$ and the decision space $\Gamma(m)$, as it is calculated for each decision policy [7]. Hence, ρ_L is given by

$$\rho_L(\mathbf{s}(m),\Gamma(m)) = \rho_L(m) = \begin{cases} \sum_{n=1}^{N} W_n(m) T_n(m), & \text{if } \alpha_n(m) > 0 \\ 0, & \text{Otherwise.} \end{cases} \tag{8.7}$$

Since ΔT is fixed, the main idea is that each time slot either supports low throughput (R_L for low bandwidth) or high throughput (R_H for high bandwidth). Therefore, the throughput of the nth CVN in the mth slot is increased by R_L if a low-bandwidth channel is allocated. Otherwise, it is increased by R_H in the case of high-bandwidth channel allocation. If no channel is allocated due to non-availability of spectrum, throughput will be zero for that particular slot. In view of this, the achieved throughput for the nth CVN in the mth slot is given by

$$R_n(m) = \begin{cases} R_L, & \text{if} \quad \gamma_n = 1,2,\ldots,L \\ R_H, & \text{if} \quad \gamma_n = L+1,\ldots,C \\ 0, & \text{Otherwise.} \end{cases} \tag{8.8}$$

The overall throughput corresponding to all CVNs, referred to as the throughput utility function ρ_R, is defined as the weighted sum of the individual transmission throughput of all CVNs with non-zero packets. Again, note that ρ_R is a function of the state vector $\mathbf{s}(m)$ and the decision space $\Gamma(m)$, as it is calculated for each decision policy. Hence, ρ_R is given by

$$\rho_R(\mathbf{s}(m),\Gamma(m)) = \rho_R(m) = \begin{cases} \sum_{n=1}^{N} W_n(m) R_n(m), & \text{if } \alpha_n(m) > 0 \\ 0, & \text{Otherwise.} \end{cases} \tag{8.9}$$

To incorporate both the latency and throughput requirements, we define the overall utility function ρ as the weighted sum of both ρ_L and ρ_R. Since ρ_L has to be minimized and ρ_R has to be maximized, a negative sign is put with ρ_R so that the overall problem becomes a minimization problem. Therefore, the joint utility function for the mth slot, $\rho(m)$, is given as,

$$\rho(m) = \beta_L \rho_L(m) - \beta_R \rho_R(m) \tag{8.10}$$

where β_L and β_R are the weights corresponding to individual utility functions ρ_L and ρ_R, respectively. Both these weights serve as design parameters to be adjusted in order to prioritize latency and throughput in a particular IoV use case or application.

The utility function given by (8.10) is valid only for the mth time slot. Assuming infinite time slots, the overall cost J of the entire transmission scheduling process can be obtained by adding the joint utility function across all the time slots as follows.

$$J = \sum_{i=0}^{\infty} \rho(\mathbf{s}(i), \Gamma(i)) \tag{8.11}$$

This formulation may be used as a foundation to work towards various optimization strategies with the aim of reducing the overall system cost J by searching for the optimal decision vector, say, $\overline{\gamma}$. An option is to use the Adam optimizer as a back-propagation neural network (BPNN) for the optimization of the traffic scheduling framework. The BPNN provides the optimized system cost, say, $\hat{J}$, that can be used in conjunction with the utility function to compute the decision vector using

$$\gamma(m) = \arg\min \rho(m) + \mu \hat{J}(m+1) \tag{8.12}$$

where $0 \leq \mu \leq 1$ is the forgetting factor to provide an option to select the weight for the future values of the utility function.

8.4 QoS-AWARE COGNITIVE COMMUNICATION

Although cognitive radio offers efficient spectrum utilization and opportunistic spectrum access beneficial for the scalability of future IoVs, it is a fundamental challenge to support the diverse QoS requirements, throughput and reliability, under rapidly varying channel conditions and availability of desired channel bandwidths. Noting that the CR is an intelligent form of SDR, a CR-based IoV with such heterogeneous QoS requirements cannot fully rely on one of the narrowband or wideband SDR waveforms. In this section, we address this challenge by presenting an adaptive communication framework for the hybrid narrowband/wideband (NBWB) waveform based

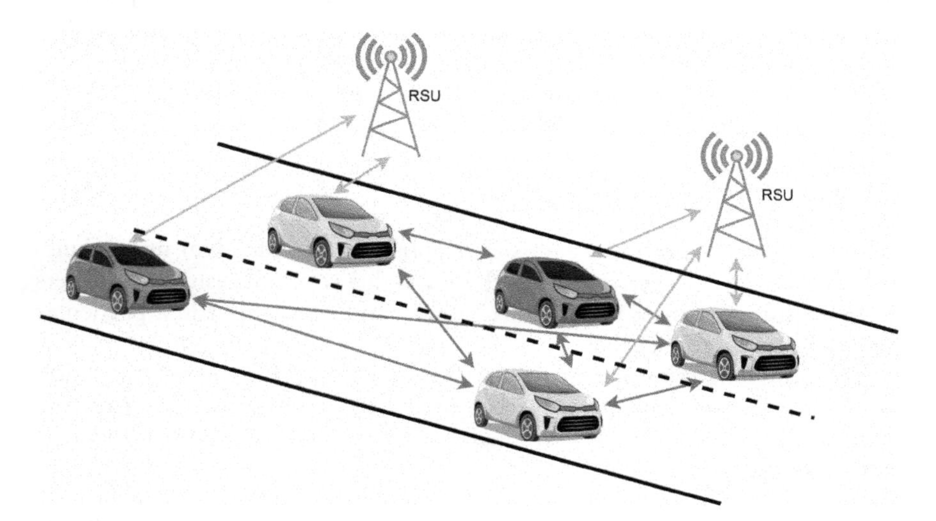

Figure 8.3 Overview of a vehicular network with a mix of short- and long-range connectivity.

on simultaneous transmission and reception of signals with multiple bandwidths through the same analog wideband front end [9]. This is facilitated by a hybrid cognitive module (HCM) that is capable of selecting appropriate waveforms with suitable system parameters.

Making a CR-based network adaptive to fulfil the varying service requirements and choosing appropriate modes of operation by adaptive modulation and coding (AMC) has been a key area of research. Extensive work is being done in this domain to make agile adaptation to varying QoS demands of each of the applications use the least resources with time and power efficiency. However, in ultra-reliable and/or time-critical environments with diverse throughput requirements, a crucial challenge is to provide connectivity to all the CVNs present in a mix of long and short ranges. It should be resolved in such a way that minimum bandwidth is utilized and end-to-end delay minimized to provide seamless communication with nodes at multiple geographical distances relative to each other. In a CR-based IoV, the CVNs present in close vicinity to each other are connected using wideband waveform and hence can communicate at higher data rates. Keeping the CVNs at longer distances connected to the network via the same wideband waveform imposes a challenge. Due to its inherent nature, the wideband waveform has less sensitivity and more absorption at longer ranges and hence cannot be used for connectivity for such distances. Narrowband waveform, on the other hand, can provide connectivity to users at longer distances but compromises the data rates. Being distinct, wideband waveforms and narrowband waveforms are dealt with differently by the front end of a CR. One of the possible solutions is to develop a hybrid NB/WB waveform front end with minimum modifications at the physical layer to incorporate the reception of multiple signals with the same or different bandwidths, communicating at the same time, using only wideband settings at the radio front end.

8.4.1 Concept of a Hybrid NB/WB Waveform

Consider an CR-based network, as shown in Figure 8.3, where some of the CVNs are in close vicinity to each other, while other nodes are at farther ranges (shown by the red arrows). The network is not completely infrastructure based, as vehicles also form a mesh network among them in addition to their connection with roadside units (RSU). The aim of the hybrid NB/WB waveform is to keep all of them connected to each other without compromising their reliability and throughput requirements. The approach works as follows. The transmitter nodes, located at different locations and having the same or different bandwidths, transmit their data simultaneously at different carrier frequencies. The analog front end of the receiver node is configured in wideband mode using wideband filter settings. The transmit carrier frequencies of the intended transmitter nodes are selected so as to pass through the analog wideband RF filter on the receiver side without leakage and overlapping. The same wideband filter settings are used at the receiver, which allows the reception of a composite signal formed by all these multi-band transmitted signals at the same time. After being converted into a digital domain, the received composite signal is processed either sequentially or in parallel. This is based on the hardware resources available. If resources are scarce, then a sequential approach is adopted for the reception of signals. A parallel processing approach is applied if latency in receiving the individual signals cannot be afforded. To separate the individual components of the signals, digital down-conversion, low-pass filters, and decoders are used [10].

Let N be the number of multi-band signals being received through the analog wideband filter. This is done while ensuring that each individual signal can occupy its required bandwidth according to its data rate requirements. Figure 8.4 shows a composite of multiple multi-band signals, where any of the individual signals can occupy any carrier frequency within the frequency range $\{f_a, f_b\}$. It is assumed that each signal has a mechanism and knowledge to avoid communications at the preoccupied frequencies. With this concept, the RF stage of the CR does not have to be switched between narrowband and wideband modes, and it will be configured in wideband mode settings with the maximum spectral bandwidth available. The wideband RF front end will receive signals of variable bandwidths ranging from the narrowest to widest possible band from the allowed pre-selected bands per the individual waveform design. To support multiple bandwidths within the proposed hybrid NB/WB waveform, suitable narrowband and wideband physical layer schemes have to be implemented. Examples of narrowband schemes include phase shift keying (PSK) and continuous phase modulation (CPM), whereas some examples of wideband schemes include orthogonal frequency division multiplexing, and filterbank multicarrier (FBMC). The more the adaptable parameters in a modulation scheme, the more flexibility there will be to achieve cognitive communication operation.

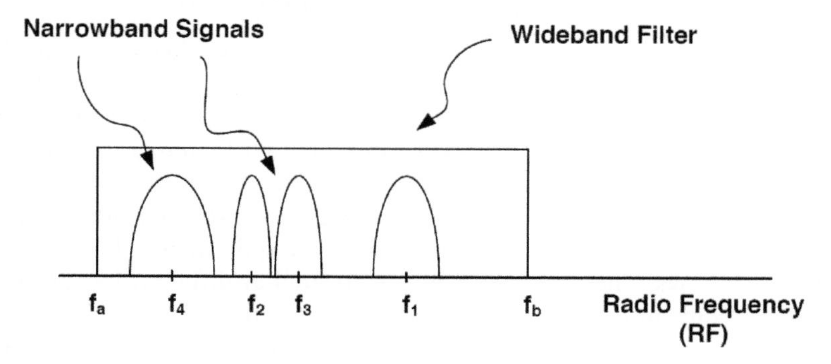

Figure 8.4 Concept of the hybrid NB/WB waveform operation. A composite of four multiband signals is being received by a single CR configured in wideband node with a wideband filter of bandwidth W (reused from our own work published in [10]).

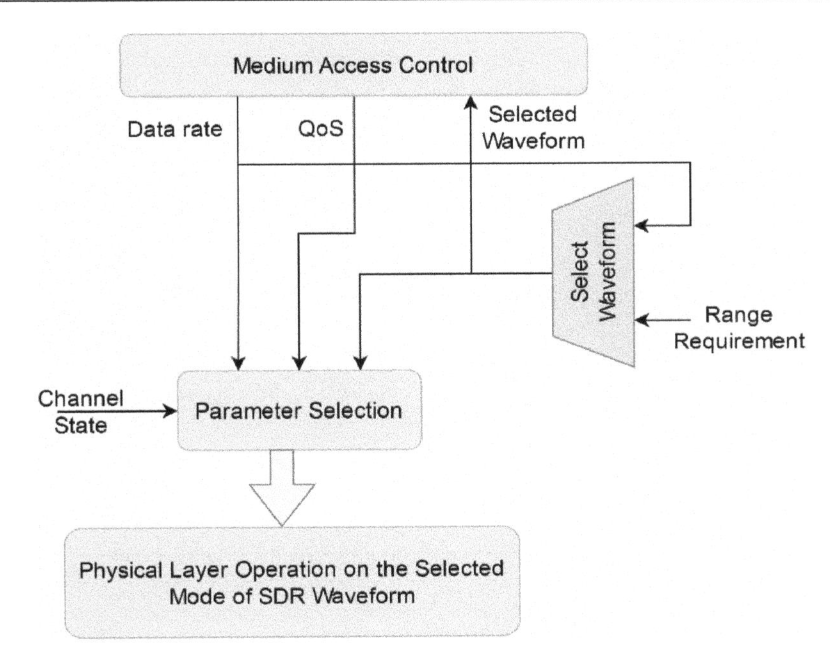

Figure 8.5 Conceptual overview of the hybrid cognitive module (HCM) [11].

8.4.2 Hybrid Cognitive Module

The successful operation of the hybrid NB/WB waveform operation also requires a cognitive mechanism to select the appropriate CR waveform with the associated parameters or mode of operation. To achieve this end, a hybrid cognitive module (HCM) as a facilitating layer of intelligence between the physical and MAC layers is illustrated in Figure 8.5. The HCM ensures the maximum utilization of the proposed PHY for addressing several service constraints. It translates the channel state based on the received signal strength indicator (RSSI) and SNR as a function of the range between CR nodes and evaluates it against the demanded data service. Based on the range and data rate requirements, the proposed HCM selects a suitable SDR waveform, narrowband or wideband. Wideband mode is selected if the operating range is small and/or the channel condition is better. On the other hand, narrowband mode is selected if the operating range is large and/or the channel condition is poor, keeping all the CR nodes connected to each other. Once an appropriate waveform is selected, it has to be configured using the optimal set of parameters based on the channel state, data rate, and QoS requirements (both coming from the MAC layer per the service needs). This selection of parameters (or mode of operation) at the physical layer of both the narrowband and wideband schemes involves the development of multiple modes of operation for both the narrowband and wideband physical

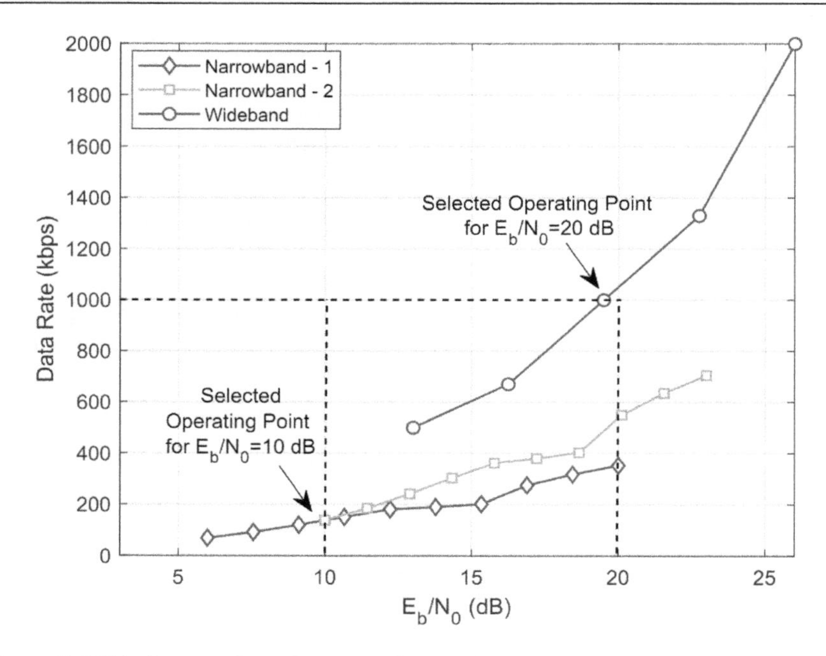

Figure 8.6 Working and evaluation of the HCM in the hybrid NB/WB waveform. Narrowband-1 and narrowband-2 waveforms have bandwidths of 100 and 200 kHz, respectively, whereas the wideband waveform has a bandwidth of 1 MHz.

layer modulation schemes with specific bit error rate (BER) performance and user throughput.

The waveform selection process using HCM is demonstrated in Figure 8.6 with the help of SNR and throughput analysis of the hybrid waveform consisting of two NB waveforms and one WB waveform. We show the theoretical throughput of all the possible modes for each waveform versus E_b/N_0, for which BER approaches zero for each mode. If the throughput requirement of a particular node is 1 Mbps and E_b/N_0 = 10 dB, HCM will select best possible mode from the 200 kHz narrowband waveform. On the other hand, if E_b/N_0 = 20 dB for the same throughput requirement, the HCM successfully finds a suitable mode of the wideband waveform, as indicated in Figure 8.6, that meets the desired throughput demand.

8.5 CONCLUSION

In this chapter, we first provided a comprehensive understanding of various IoV application scenarios along with their diverse QoS requirements. It is highlighted that the key IoV use cases impose heterogeneous requirements on latency, throughput, and reliability. To provide a solution to this challenge, a priority-aware classification of various IoV use cases is initially provided to help in developing transmission scheduling algorithms. Afterwards, we presented a data traffic scheduling framework aiming to maximize the overall throughput and minimize the overall latency by formulating a hybrid utility function. We also presented a cognitive communication framework to ensure the connectivity of all the nodes present in a mix of long- and short-range networks by simultaneously meeting the QoS and throughput requirements under varying channel conditions. In this context, a concept of hybrid narrowband/wideband waveform is presented, along with the hybrid cognitive module as a facilitating layer between the physical and MAC layers.

REFERENCES

1. Liu, K., Xu, X., Chen, M., Liu, B., Wu, L., and Lee, V. C., A hierarchical architecture for the future Internet of Vehicles. *IEEE Communications Magazine*, *57*(7), pp. 41–47, 2019.
2. Eze, J., Zhang, S., Liu, E., Chinedum, E. E., and Hong, Q. Y., Cognitive radio aided Internet of Vehicles (IoVs) for improved spectrum resource allocation. In *2015 IEEE International Conference on Computer and Information Technology; Ubiquitous Computing and Communications; Dependable, Autonomic and Secure Computing; Pervasive Intelligence and Computing*, pp. 2346–2352.2015.
3. Hussain, S. A., Yusof, K. M., Hussain, S. M., and Singh, A. V., A review of quality-of-service issues in Internet of Vehicles (IoV). In *2019 Amity International Conference on Artificial Intelligence (AICAI)*, pp. 380–383.2019.
4. "3GPP TR 22.886 v15.0.0, Technical specification group services and system aspects; study on enhancement of 3GPP support to 5G V2X services. Release 15," December 2016.
5. Kovács, G. A., and Bokor, L., Integrating artery and Simu5G: A mobile edge computing use case for collective perception-based V2X safety applications. In *2022 45th International Conference on Telecommunications and Signal Processing (TSP)*, pp. 360–366.2022.
6. Chen, S., Hu, J., Shi, Y., Zhao, L., and Li, W., A vision of C-V2X: Technologies, field testing, and challenges with Chinese development. *IEEE Internet of Things Journal*, *7*(5), pp. 3872–3881, 2020.
7. Yu, R., Zhong, W., Xie, S., Zhang, Y., and Zhang, Y., QoS differential scheduling in cognitive-radio-based smart grid networks: An adaptive dynamic programming approach. *IEEE Transactions on Neural Networks and Learning Systems*, *27*(2), pp. 435–443, 2015.
8. Khan, M. W., Zeeshan, M., Farid, A., and Usman, M., QoS-aware traffic scheduling framework in cognitive radio based smart grids using multi-objective optimization of latency and throughput. *Ad Hoc Networks*, *97*, p. 102020, 2020.
9. Shahzad, K., Farooq, M. U., Zeeshan, M., and Khan, S. A., Adaptive multi-input medium access control (AMI-MAC) design using physical layer cognition for tactical SDR networks. *IEEE Access*, *9*, pp. 58364–58377, 2021.

10. Shahzad, K., Gulzar, S., Zeeshan, M., and Khan, S. A., A novel hybrid narrowband/wideband networking waveform physical layer for multiuser multiband transmission and reception in software defined radio. *Physical Communication*, *36*, p. 100790, 2019.
11. Zeeshan, M., Shahzad, K., and Farooq, M. U., NOMA-enabled cognitive communication based on hybrid narrowband/wideband SDR waveform. In *3rd International Informatics and Software Engineering Conference*, 2022.

Index

For Product Safety Concerns and Information please contact our EU
representative GPSR@taylorandfrancis.com
Taylor & Francis Verlag GmbH, Kaufingerstraße 24, 80331 München, Germany

www.ingramcontent.com/pod-product-compliance
Ingram Content Group UK Ltd.
Pitfield, Milton Keynes, MK11 3LW, UK
UKHW021120180425
457613UK00005B/167

* 9 7 8 1 0 3 2 2 5 7 5 5 6 *